INTELLIGENCE: IN
THE BATTLE FOR
THE MIND

INTELLIGENCE: THE BATTLE FOR THE MIND

H. J. EYSENCK _versus_ LEON KAMIN

PAN BOOKS LONDON AND SYDNEY

**This book was devised and produced
by Multimedia Publications Inc**

First published in Great Britain 1981 simultaneously by
Pan Books Cavaye Place, London SW10 9PG, in paperback
and by **The Macmillan Press Ltd** London and Basingstoke, in hardcover

Editor : Susana Raby

**Production : Arnon Orbach
Design/Layout : Mike Spike**

ISBN 0 330 36399 4 (paperback)
ISBN 0 333 31279 1 (hardcover)

British Library Cataloguing in Publication Data
Eysenck, Hans
Kamin, Leon

Intelligence: The battle for the mind
References pg. 171–172, 182–187.
Includes index.

Typeset by CCC, printed and bound by William Clowes (Beccles) Limited,
Beccles and London, England

Photo Credits: Rex Features Ltd—85, 125, 141, 143.
Applied Psychology Unit (MRC) Cambridge—25.
Miki Koren—12, 58, 86bc, 92, 109. Israel Sun Ltd—48, 86a, 135, 151.
Multimedia Publications Inc have endeavoured to observe the legal requirements with regard
to the rights of the suppliers of graphic and photographic materials.

CONTENTS

INTRODUCTION

INTELLIGENCE—one of the most important ways by which we judge one another—is a powerful and emotive issue for parents, teachers, employers and even politicians. But what exactly is intelligence? How is it formed? How much is it related to hereditary factors, and how much to social ones? And, most important of all, can we develop an objective, scientific way of measuring this aspect of ourselves?

This very loaded word has been at the centre of controversial, and at times bitter, debate for many years. Few subjects in the social sciences or humanities have inflamed passions with such ferocity, and many scientists, politicians and journalists are guilty of mud-slinging. Their claims and counter-claims have created a quagmire.

It is still difficult to find a majority of experts anywhere who would agree on an acceptable definition of intelligence, as well as on its implications for human behaviour. Professor Eysenck seeks to persuade us that genetic factors determine not only our intelligence but many other aspects of behaviour as well. Professor Kamin argues with equal determination that intelligence is shaped primarily by environmental factors. This book presents a great debate between two well-known advocates holding diametrically opposed views on intelligence. It gives the reader the rare opportunity of weighing up their arguments, which are sharp, uncompromising and controversial.

Professors Eysenck and Kamin agreed not to see each other's manuscripts while they were being written. After the finished manuscripts had been accepted for publication, each was sent the other's (previously unseen) manuscript and invited to write a rejoinder. It was understood that the original manuscripts could not be altered in the light of the rejoinders. These rejoinders form part of the book and round off a remarkable clash.

Criticism and debate are the life-blood of science, and in this very active area of such popular concern, it seems only right that both sides should outline their case, and be subject to informed criticism. Perhaps we should set the stage with the words of the authors themselves:

"The concept of intelligence, and the question of its heritability, both have important psychological and social consequences. On this, as well as on several substantive points, we are both agreed. There are also, however, a number of points on which we are disagreed; fortunately most if not all of these are subject to scientific enquiry of an empirical nature, and the results of the many hundreds of studies in the field are discussed in this book in an effort to come to some agreement, or, if that should prove impossible, to delineate as clearly as possible the areas of disagreement that remain, and the possible ways in which these disagreements could be resolved."

(H. J. EYSENCK)

"The publisher of this volume has asked both Professor Eysenck and me to explain briefly how this book came about. The format of a 'debate', once it was proposed, seemed to me entirely appropriate. My purpose has not been to try to change Professor Eysenck's mind; of that I despair. But his has been a voice of considerable public influence. I do not want it thought that his opinions represent those of all scientists, or of all psychologists. Nor would I want his opinionated views to be thought of as scientific facts. I think that he is wrong, and I think that the facts demonstrate this. I hope and believe that in the process of rational debate I can convince readers that this is so."

(LEON KAMIN)

H. J. Eysenck was born in Berlin, Germany, in 1916. He left Berlin in 1934 in protest against the Hitler movement, and studied language and literature for a while in Dijon, France, and Exeter, England, before taking up psychology at University College, London, under Cyril Burt. After obtaining his Ph.D. there, he joined the Mill Hill Emergency Hospital during the war as research psychologist, and after the war became psychologist to the Maudsley Hospital. Later he founded the Psychological Department and Laboratory at the Institute of Psychiatry, which is associated with the Maudsley Hospital and is part of the University of London. He was appointed Reader and then Professor at the University of London, and still runs the Department, which has grown to number some thirty academic staff. He has published some three dozen books and some six hundred scientific articles. His main academic interests are personality and individual differences, intelligence, behaviour therapy, behavioural genetics, the study of social attitudes and experimental aesthetics.

Leon Kamin was born in Taunton, Massachusetts on 29th December, 1927. He is currently the Dorman T. Warren Professor of Psychology at Princeton University, where he was chairman of the Department of Psychology from 1968 to 1974. Leon Kamin is a Fellow of the American Psychological Association, and a member of various professional psychology associations. He is a past President of the Eastern Psychological Association and is currently an executive committee member of the Division of Experimental Psychology of the American Psychological Association. He received the Martin Luther King Junior Award from the New York Society of Clinical Psychologists in 1976, and a special award of the National Education Association Committee on Human Relations in 1978. Professor Kamin has reviewed numerous books, has published over fifty scientific articles, and has written chapters in many books. He is also the author of *The Science and Politics of IQ*.

I
WHERE DOES THE CONCEPT COME FROM?

Intro

The man in the street often speaks of "intelligence". So does the professional psychologist. The meanings attached to the term are not always identical, and indeed may at times seem contradictory. Nevertheless, there will be general agreement that whatever "intelligence" may be, it is not a *thing*, like a table or a chair, or a pig, but a *concept*, a term which carries meaning and can only be understood by virtue of a whole set of facts and theories associated with it.

In the heat of the discussion about intelligence, its inheritance and its social implications, this is sometimes lost sight of. But, as we shall see, the fact that intelligence is a concept is of vital importance in trying to understand just what it means, what its limitations are, how it can be defined and measured, and whether or not it is inherited. The position taken in this book is that intelligence as a scientific concept is precisely analogous to temperature and other scientific concepts, and that the difficulties its measurement gives rise to are no different from those to which the measurement of temperature and other scientific concepts gives rise.

THE ANCIENT GREEK CONTRIBUTION

The origins of the concept are lost in antiquity. We know that Plato and Aristotle already drew a distinction between the *cognitive* aspects of human nature (those concerned with thinking, problem solving, meditating, reasoning, reflecting and so on) and the *hormic* aspects of human behaviour (those concerned with emotions, feelings, passions and the will). Cicero later coined the term *intelligence*. We still use the term intelligence to refer to a person's cognitive powers and intellectual abilities.

Having created the concept of intelligence, the Greeks went on to make other important contributions. Aristotle contrasted the *observed activity* or behaviour of a person with some hypothetical *underlying capacity* or *ability* on which it depended. The concept of ability is sometimes called a "latent structure concept": we postulate some latent or underlying structure to account for the ability we have observed. Intelligence is just such a latent structure concept. It has to be deduced from observed behaviour using the rules of scientific experimental procedure; and we postulate some underlying structure in the nervous system to account for intelligent behaviour.

The nature–nurture distinction

Plato contributed the distinction between *nature* and *nurture*, and clearly favoured genetic causes in accounting for individual differences in intellect and personality. Many readers will be familiar with his famous fable of the different metals: "The God who created you has put different metals into your composition—gold into those who are fit to be rulers, silver into those who are to act as their executives, and a mixture of iron and brass into those whose task it will be to cultivate the soil or manufacture goods." He also recognised the fact of *genetic regression* (the tendency of very intelligent or very dull parents to have children who regress to the mean, in other words who are less bright, or less dull, than their parents): "Yet occasionally a golden parent may beget a silver child, or a silver parent a child of gold; indeed, any kind of parent may at times give birth to any kind of child."

The odds against a black father and a white mother producing a strikingly fair-haired baby are fairly high.

Plato considered it the most important task of the Republic to allocate tasks and duties according to the innate abilities of the person concerned: "That first and foremost they shall scrutinise each child to see what metal has gone to his making, and then allocate or promote him accordingly." The penalty for failure should be severe, "for an oracle has predicted that our state will be doomed to disaster as soon as its guardianship falls into the hands of men of baser metal." Modern meritocratic society has come close to fulfilling at least some of Plato's dreams by promoting men of intelligence, though intelligence was not the only quality which distinguished men of gold from those of silver or those of iron and brass.

MODERN DEVELOPMENTS

In the last century, the notion of intelligence was taken up by the philosopher Herbert Spencer, by the statistician Karl Pearson, and by Darwin's cousin, the all-round genius Sir Francis Galton. They introduced to the study of intelligence the notions of measurement, evolution, and experimental genetics. To these contributions should be added those of the physiologists, particularly the clinical work of Hughlings Jackson, the experimental investigations of Sherrington, and the microscopic studies of the brain carried out by Campbell, Brodman and others. This physiological work did much to confirm Spencer's theory of a "hierarchy of neural functions" in which a basic type of activity develops by fairly definite stages into higher and more specialised forms. The brain, it was found, always acts as a whole. Its activity, in Sherrington's words, is "patterned, not indifferently diffuse", and the patterning itself "always involves and implies integration". Lashley later contributed the concept of "mass action" of the brain, which states that cognitive functioning is governed by broad areas of the brain rather than specialised small areas. Mass action was theoretically identified with intelligence by several writers.

Spearman's "g": an all-embracing mental ability

The person who fused all these different notions into a proper psychological theory was Charles Spearman, for many years Professor of Psychology at University College, London. He started with a very simple idea which proved to be exceedingly fruitful. He argued that if there existed some all-round, all-embracing cognitive ability which enabled a person to reason well, solve problems and generally do well in the cognitive field—Spearman called it "g"—then it should be possible to construct a large number of different problems, of varying difficulty, to put this ability to the test.

At around the same time, Alfred Binet in France and Hermann Ebbinghaus in Germany were in fact devising such tests; what Spearman added was a rather simple statistical idea. Put briefly, it was that it should be possible to show whether some people are better at all types of cognitive tests than others—as the very notion of intelligence would imply—simply by giving large numbers of tests to a random sample of people and comparing the results of the tests or test items by a process known as correlation. If the hypothesis is true, then all the correlations should be positive. In other words, being good at one kind of test would make you likely to be good at other types. (A correlation is simply a statistical device for showing the degree to which two factors are related and is expressed as a figure ranging from 0 to 1. A positive correlation of 1.00 indicates a perfect correspondence; 0.00 indicates the absence of any relationship at all. A negative correlation, which is expressed as, say, -0.75, indicates that the two factors being compared *are* related but inversely: the higher the one, the lower the other.)

Hundreds of studies have since shown that Spearman was right: cognitive tests of any kind correlate positively when the tests are carried out on people chosen at random from the population. Spearman, however, went one step further. He showed mathematically that if ability at a given cognitive task is broken down into two distinct elements which are examined separately—the first being general cognitive ability or intelligence, and the second being the specific ability to perform that particular task—the pattern of correlations between tests assumes a very specific form. Intercorrelations between different tests are expressed in the form of a rectangular table or grid mathematicians call a *matrix*. The particular pattern Spearman found is known as a "matrix of rank 1", which would be very unlikely to occur by chance. He concluded that by and large the theory was supported. We shall see in a moment to what extent we can still accept this conclusion; let us here merely note that it represents a complete break with the past because now we have a theory which gives rise to testable, quantifiable hypotheses; this distinguishes it from the theories of Plato, Aristotle, Spencer and all the others.

Primary abilities—Thurstone's blast . . .

The first one to test Spearman's theory on a large scale, and to claim that he had disproved it, was Professor LL Thurstone of the University of Chicago. Using 56 tests of various intellectual abilities on large numbers of University of Chicago graduates, calculating the correlations between them and analysing them according to the rules of matrix algebra, he concluded that Spearman was wrong: his correlations, which he claimed demonstrated the presence of a general cognitive ability, were in fact measuring a number of different so-called "primary abilities", such as verbal ability, numerical ability, visuo-spatial ability, memory, and so on. This finding seemed to agree well with the earlier speculations of Alfred Binet, the French psychologist who devised the first widely accepted test of intelligence, and who believed that intelligence was made up of a number of different mental "faculties" which were being tested by different components of the tests. (Actually Binet's theories are not easy to unravel, as he also persisted in thinking of his test as measuring some central faculty of "intelligence".)

. . . and Spearman's counterblast

Spearman did not accept Thurstone's results, for two main reasons. In the first place, Thurstone had only tested highly intelligent and specially selected students. His subjects did not constitute a random sample of the population—the range of intelligence in his sample was severely restricted. This is crucial; you would not expect to be able to make pronouncements about the height of the average Englishman if you included in your analysis only the heights of pre-war London policemen, who were required to be over six feet tall. Restriction of range was not sufficient to eliminate the positive correlations between all the tests, but it clearly reduced them considerably.

The other objection Spearman raised was that he had specified in presenting his theory that the tests should not be too similar; if they were very similar, then the specific factors would overlap and produce irrelevant correlations. Many of Thurstone's tests were rather similar; for instance, he had several different vocabulary tests which obviously measured pretty much the same ability. The correlations between them were therefore due not only to a general factor of intelligence but also to the fact that specific abilities were being measured more than once, confusing the issue.

A paradigm emerges

Thurstone, like the good scientist he was, repeated his study, with his wife Thelma, on a large group of unselected schoolboys, thus overcoming the criticism that he had worked only with uniformly intelligent students. When he did this, he found that there were a number of what he persisted in calling "primary factors". These, however, correlated highly with each other. When he worked out the correlations between his primary factors, the matrix, or grid, they formed was very close to being a matrix of rank 1—which is what Spearman's theory demanded. He concluded that the tests did measure something very similar to Spearman's general intelligence, or "g", but that they also measured a number of primary abilities, over and above intelligence, and independent of it. Spearman and his students had also by now found evidence for various factors such as verbal and numerical ability. Consequently, final agreement was reached on a paradigm which has lasted to this day. The paradigm states that different people have different abilities for solving intellectual problems, and that particularly important among these abilities is general intelligence. There are also specific abilities to deal with specific types of problems—for instance, verbal, numerical, visuo-spatial, mechanical or memory abilities—which can be very important under special circumstances. In addition, every test has its own unique contribution attached to it which interferes with the measurement of intelligence or special abilities. This error can be eliminated by using many different tests incorporating as many different kinds of material as possible.

There have been many criticisms of this paradigm, and alternative theories have emerged. I shall argue that though some of the criticisms have been well taken, none has been able to shake the paradigm in any serious way. Alternative theories, such as those of Guilford and others, have failed to make their case, and have been shown to be faulty in important respects. This chapter has introduced the paradigm briefly; the following chapters will discuss various aspects of it in detail.

2

WHAT ARE INTELLIGENCE TESTS?

Usually, intelligence tests are made up of a variety of items to test the specific mental abilities believed to play a part in general cognitive ability. Items are usually arranged in ascending order of difficulty, with dissimilar ones juxtaposed to increase interest.

The first actual scale for the measurement of intelligence was produced by Binet in Paris, for the purpose of testing children in school. It was based on the concept of mental age introduced by SE Chaillé in 1887, who calculated a child's mental age from the level of difficulty of the cognitive problems he could solve.

THE CONCEPT OF MENTAL AGE

The difficulty level of a problem was established by discovering the average age at which most children could solve it. Thus if a three-year-old succeeded at problems usually solved by four-year-olds, his mental age would be four and his chronological age three. Conversely, if at a chronological age of ten he only succeeded with problems typically solved by an eight-year-old, and failed at the nine-year-old level, his mental age would be eight. These two concepts were later put together in the form of the so-called intelligence quotient:

$$IQ = \frac{MA}{CA} \times 100.$$

MA stands for mental age, and CA stands for chronological age. The 100 is introduced to get rid of the decimal point. Bright children have IQs over 100, dull children under 100, and the exactly average child has an IQ of 100.

Figure 1 shows the kind of IQ distribution we find in the population in general and gives an indication of the meaning of different IQs. The terms are purely descriptive, of course, but are useful as a rough guide.

It may be of interest to look at some of the test items Binet used in his 1908 intelligence scale. At the age of three a child can point to nose, eyes

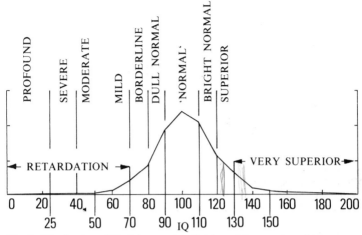

Fig. 1. Distribution of IQ giving rough indication of the meaning of scores

or mouth; can repeat sentences of six syllables; can repeat two digits; can enumerate objects in a picture and give his family name. At the age of four he knows his sex, he can name certain objects shown to him, such as a key, pocket knife or penny; he can repeat three digits and can indicate which of two lines, 5cm and 6cm in length respectively, is the longer.

At the age of five, the child can indicate the heavier of two cubes, one weighing 3 grammes and the other 12 grammes; he can copy a square, using pen and ink; he can construct a rectangle from two pieces of cardboard, having a model to look at; and he can count four pennies. At the age of six he knows right and left as shown by indicating right hand and left ear; he can repeat sentences of 16 syllables; he can define similar objects in terms of their use; he can execute a triple order; he knows his age, and he knows morning and afternoon. At the age of seven he can tell what is missing in an unfinished picture; he knows the number of fingers on each hand, or both hands, without counting them; he can copy a diamond, using pen and ink; he can repeat five digits; he can describe pictures as seen; he can count 13 pennies; he knows the names of four common coins.

These are typical of the accomplishments of younger children. While the facts of development were of course known in broad outline, it was crucial for the construction of Binet's scale to determine exactly the average age at which the child becomes able to carry out various tests. Later workers such as Piaget have followed Binet in describing stages of development; tests used by Piaget correlate very well with Binet's.

DEVISING TEST ITEMS

Frequently nowadays, IQ tests are not individual tests administered by psychologists but group tests given to many people at the same time. To make scoring easier, the subject is asked to select the correct answer from the several alternatives presented. Figure 2 shows typical items used in a group test.

Items 1 and 2 are *series* problems, respectively letter series and number series. Items 3, 7 and 8 are different types of *matrix* problems. Item 4 is an *incomplete sentence* problem. Item 5 is a *relations* problem. Item 6 is a *dominoes* problem. There are many more types of problem, but these are sufficient to give an idea of what IQ tests are like.

How are such items devised? There are several major rules. The first is that the item should not take too long to solve; we have only a limited

(1)   A   C   F   J   O   U   Complete.

(2)   3   8   12   15   17   Complete.

(3) Select the correct figure from the six numbered ones

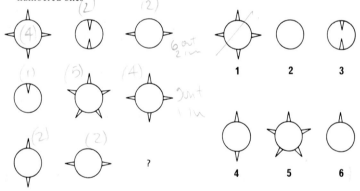

(4) The dog _____ loudly at the stranger.   Complete.

(5) | : —— = ○ : ○ ⬭ ⊗ ——   Underline right answer.
                              1   2   3   4

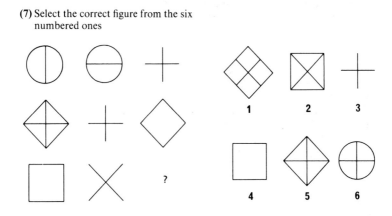

(6) Complete

(7) Select the correct figure from the six numbered ones

(8) Select the correct figure from the six numbered ones

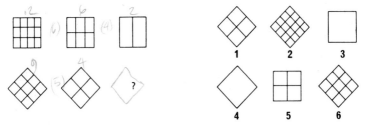

Fig. 2. Typical group test items

period for testing in school, or in the army, or in industry, and many test items are needed to get a realistic idea of a person's abilities. In the second place, items must be so devised as to have a single correct answer. Thirdly, the test should not be one of knowledge but of problem-solving; in other words, all the elements in test items should be equally known or equally unknown to all the children or adults taking part. This may be difficult to achieve when very dissimilar populations are being tested, but it can be approximated very closely in relatively homogeneous populations, where education is compulsory and all children go to school. Even then, items requiring advanced knowledge of any subject must of course be avoided.

Novel content

Above all, items of an intelligence test should follow the laws of *noegenesis* as originally formulated by Spearman. Noegenesis means the production of new or novel content, based on the relations observed between the elements of a given problem; the major rules of interest here are the *eduction of relations* and the *eduction of correlates*. The former is illustrated in Figure 3a, the latter in Figure 3b. Given two fundamental elements, or "fundaments"—for instance the words "black" and "white"—we can educe the relation: opposites. Given the fundament, "black", and the relation, opposites, we can educe the correlate: "white". Thus from known fundaments and relations, we can educe new material implicit in the problem. As an example, consider the matrix problem below. There are various relations between the figures shown: for instance, gradations of black, grey and white; shapes (square, round, triangular), and signs on top of the major figures ($+$, C, T). In each row, and in each column, there is one example of each, and the various relations between the fundaments enable us to deduce that the missing figure is number 6. The process of arriving at this conclusion is noegenetic; the final decision has to be arrived at by a cognitive process, or series of processes, basic to all cognitive problem-solving. The problem is, of course, an easy one, but young children or persons of low IQ will nevertheless have difficulties with it, or even be unable to solve it.

In the construction of a test, a number of items are selected according to principles to be discussed in a later chapter, put together in a test and administered to large samples of the population. The results make it possible to *standardise* the test: knowledge is derived from them about the level of difficulty of each item, the age at which the item is typically solved by the average youngster, any differences between the sexes in ability to solve a particular problem, and so on.

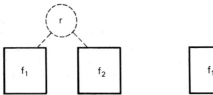

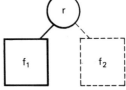

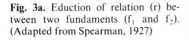

Fig. 3a. Eduction of relation (r) between two fundaments (f_1 and f_2). (Adapted from Spearman, 1927)

Fig. 3b. Eduction of correlate (f_2) from fundament (f_1) and relation (r). (Adapted from Spearman, 1927)

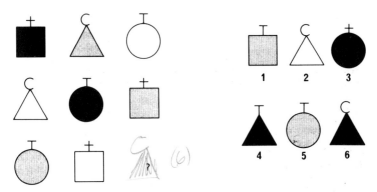

Fig. 4. Eduction of relations and correlates, a typical test item of the Matrices type

The prime reason for regarding such tests as tests of intelligence is the simple observation that children clearly grow more intelligent in an absolute sense as they grow older; the average ten-year-old is brighter than the average four-year-old. Thus mental age is an index of mental ability, and in relation to chronological age it gives us some indication of the degree to which a child is advanced or retarded. This was the original concept on which mental testing was based, and it still seems a pretty solid one nowadays, although by now we have many more ways of supporting the view that IQ tests measure intelligence.

USES OF THE TERM "INTELLIGENCE"

It is important to distinguish between different uses of the term intelligence. DO Hebb has suggested the use of the terms *intelligence A* and *intelligence B*. Intelligence A is the basic potentiality of the organism to learn and to adapt to its environment; it is determined by the complexity and plasticity of the central nervous system, which in turn is determined by the genes. Some people are better endowed with these genes and therefore have greater potential for mental development. This development does not take place in a vacuum, of course, but depends on suitable stimulation from the physical and social environment in which the child is reared.

Intelligence B is the level of ability a person actually shows in behaviour. This, of course, is not genetic, nor is it simply learned or acquired. It is a product of the interplay between nature and nurture, between genetic potential and environmental stimulation. A third definition of intelligence—intelligence C—might also be introduced to refer to the actual measurement of intelligence B by IQ tests. Clearly, IQ tests will only partially measure intelligence B, and will not be able to encompass the whole of it. These distinctions are interesting and important, and the evidence we will discuss enables us to come to some conclusions about their relationship. Intelligence C—that is, IQ—is pretty closely related to intelligence B, and the evidence suggests that intelligence A is pretty closely related to intelligence B in our type of society. We will later on see the reasons for believing these conclusions to be true. We will also see that quite recently methods have been developed to assess intelligence A directly, and that the measurements they give rise to show a close relationship with scores obtained on IQ tests.

Cultural factors and crystallised ability

The degree to which environmental and cultural factors affect items in an intelligence test is variable. Items of the simple series type, the matrix type or the dominoes type are clearly very little affected by cultural factors, whereas items involving language, particularly vocabulary test items, are much more so. At one extreme we have what are sometimes called tests of *fluid intelligence*—culture-free or *culture-fair* tests minimally dependent on knowledge, education or cultural factors. At the other extreme we have tests of *crystallised intelligence* which draw on knowledge and information more likely to have been acquired by intelligent persons than dull ones. Where the acquisition of knowledge is reasonably standard, the amount of knowledge acquired might be considered a direct measure of intelligence. Strictly speaking by intelligence the psychologist normally means only fluid intelligence, but in countries where the educational system is reasonably egalitarian, crystallised intelligence may appear very similar to fluid intelligence—certainly in the United States of America, the United Kingdom and in continental Europe, the two correlate quite highly.

This section has been largely descriptive; we have raised certain questions about the meaning of intelligence and its inheritance, and about the difference between fluid and crystallised ability, but we have not attempted to answer them. We will try to give the answers in subsequent sections.

3

WHAT DO INTELLIGENCE TESTS MEASURE?

The last chapter looked at typical examples of intelligence test items. To understand what IQ tests measure, it is crucial to understand how intelligence tests made up of these items are put together. Critics often give the impression that psychologists make up the tests in a quite arbitrary fashion, selecting items they prefer, for some inscrutable or not so inscrutable reasons of their own. Thus it is sometimes alleged that white, middle-class psychologists pick out items that favour white, middle-class children.

We shall see later whether the tests constructed by psychologists do in fact favour white, middle-class children; in the meantime, it is interesting to look at how such tests are in fact constructed. The method used will be shown to be highly objective in nature, contrary to critical objections.

CONSTRUCTING IQ TESTS

Let us begin with two facts which are not in dispute. The first is that all tests of intelligence correlate positively together, a fact sometimes known by the name "positive manifold". This means that if we took a random sample of the thousands of tests that can be or have been constructed, it would correlate very highly with another random sample of test items, and the larger the two samples, the more perfect would be the correlations. No conscious choice would be involved at all; any appropriate items or tests would do to make up this hypothetical super-test. In actual fact, of course, we cannot construct a test with an infinite number of test items, and it would be a waste of time and energy to make a random selection. We must therefore take into account the second widely acknowledged fact.

This second fact is that test items are of many different kinds and can be categorised, as was indicated in the last chapter. A good test of intelligence should obviously include as many different types of item as possible; it should not be made up exclusively of items of one particular type, or related to one particular ability. The more varied it is the better, which is what made the original Binet test so valuable—it included a great variety of different items covering all the different primary abilities later recognised by Thurstone, including verbal, numerical and visuo-spatial abilities.

We now have two principles. The first one states that as long as we have a large enough number, almost any cognitive test items will do (Spearman called this principle "the indifference of the indicator"). The second is that test items should be as varied as possible in order to include all the different aspects of intellectual functioning, and not to overstress one particular primary ability. However, to these should be added a third principle, namely that of preferring "good" items to "bad" ones.

What makes a test item good or bad?

What makes an item good or bad? The answer given can be either theoretical or empirical; preferably it should be both.

On the theoretical level, we have Spearman's principle of noegenesis. (Noegenesis, remember, is the stimulation of novel thinking.) A good item will embody the principle of noegenesis, a bad item will not. Thus an item like "*Carmen* is to *Bohème* as Bizet is to: Verdi/Puccini/Wagner/Strauss" is bad because it depends almost entirely on acquired knowledge, not on any kind of noegenesis; either you know that Bizet wrote *Carmen* and that Puccini wrote *La Bohème* or you don't. It is, of course, more likely that an intelligent person will know this than a dull one, but this knowledge would be an extreme example of crystallised ability and not at all suitable for the measurement of general intellectual ability. This theoretical criterion should be supplemented by an empirical one.

On the empirical level, Spearman argued—and it is universally recognised that his argument was correct—that while all cognitive tasks involve intelligence for their solution, they do so to unequal degrees. Some tests are better than others in the degree to which they involve general cognitive ability for their solution. Can we discover that degree? The answer is of course yes. If all cognitive tests measure "g", but to different degrees, then good tests should correlate more highly with all the other tests than should bad tests. Examination of how large numbers of tests or test items correlate with each other, followed by a more technical type of analysis called *factor analysis*, should tell us objectively which items are good and which items are bad. Ideally, items should only be admitted if they comply to both the theoretical and the empirical principles, although not all test constructors have followed them, and some tests are of a low technical standard.

Other requirements

A number of other conditions have to be fulfilled in the construction of a proper test of intelligence. To take one example, it must include items of different levels of difficulty; tests made up only of easy or only of difficult items obviously cannot discriminate between subjects, The level of difficulty of an item can be measured by administering it to large groups of people and seeing what percentage can solve the problem adequately within a given period of time.

This is not the place to go into all the other requirements of IQ tests; let us merely note that the choice of test items is not left to the subjective whims of the experimenter. If he wishes his test to be widely accepted, he must follow certain objective procedures which ensure that whatever the subject's social class or skin colour, the outcome will be pretty much the same.

IQ tests are made up, then, of a large number of individual items, differing in difficulty level, differing in the specific abilities needed to solve them, and all requiring a reasonable degree of "g" to be solved successfully. What do these tests measure? There are essentially two ways of answering this question. The first relates to the *internal validity* of the tests—their agreement with each other—the second to their *external validity*. We shall deal with internal validity in this chapter and with external validity in the next.

INTERNAL VALIDITY

Intelligence is what IQ tests measure

Psychologists, when asked what intelligence is, sometimes say, with tongue only partly in cheek, that it is what intelligence tests measure. This often produces amusement among listeners not trained in science, for it seems to be nothing more than a tautology. However, in science definitions of this kind—so-called *operational definitions*—are quite common; indeed many scientists believe they are the only kind of scientific definition which is acceptable. You define a concept in terms of the ways in which you measure it and the measurements achieved. This is not tautological because the measurements are derived from a theory and can be used to verify or invalidate it. The statement that intelligence is what IQ tests measure is not circular because it stands to be disproved by IQ measurements themselves. Thus if we found that our tests of intelligence did not all correlate positively with each other, we would have to conclude that they did not measure intelligence. We would say that they lacked internal validity.

If we are going to define a concept by the tests which measure it, it is

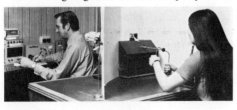

Two test situations, both designed to test attention to subsidiary tasks.

obviously crucial that the tests should have internal validity—that they should agree with each other. When we say that a test has internal validity, we are saying that it measures a factor objectively, with a degree of error which can itself be measured, and that it correlates positively with other tests of the same factor. "G" is just such a factor, though we cannot at this stage say that "g" is the same as intelligence as the term is understood by the man in the street.

The analogy with heat

On the subject of operational definition, it is helpful to draw a comparison between intelligence and heat. Heat has been measurable since Torricelli constructed the first thermoscope some 300 years ago. But do we have an adequate definition of heat, except that it is what is being measured by our thermometers? A study of physics shows that we do not. There is no single theory of heat, but two rather different theories, the thermodynamic and the kinetic.

Thermodynamics deals with abstract concepts of a purely quantitative kind: temperature, measured on a thermometer; pressure, measured as a force exerted per unit area; and volume, measured by the size of the container. *Nothing is said in the laws of thermodynamics about the nature of heat.* On the other hand, the kinetic theory of heat, which goes back to Bernouilli and his famous treatise on hydraulics, attributes differences in heat to the motion of small particles of which all bodies and fluids are made up; the faster the particles move, the hotter the body. This is a nice theory giving a picture of events which is readily visualised. But, even today, many phenomena which accord easily with thermodynamic theory are not amenable to kinetic interpretation. There is no unified theory of heat, and ultimately heat is defined in terms of the measuring instruments used, very much as intelligence is.

Different tools for different needs

Surely, the reader may object, different types of intelligence tests are used for different purposes; can they all be said to measure the same quality? But exactly the same is true of thermometers. Different types of thermometer are used for different temperature ranges. Mercury freezes at $-39°C$ and boils, under atmospheric pressure, at $357°C$, although it can be made to serve up to about $550°C$ by filling the space above the liquid with nitrogen, which is compressed as the mercury expands, and raises its boiling point. Alcohol thermometers can be used at lower temperatures; ethyl alcohol boils at $78°C$ and freezes at $-115°C$ and is preferred for measurements in polar regions.

High temperatures are usually measured by observing the radiation from a hot body—a technique called pyrometry. Pyrometers, whether they are of the total radiation type or the optical type, cover a different temperature range to other instruments. Resistance thermometers, which make use of different physical properties again, have their own disadvantages. Then we have the constant-volume gas thermometer, which depends on the reactance of the welded junction of two fine wires, as well as many others.

Furthermore, different ways of measuring temperature do not give the same results. When a mercury-in-glass thermometer reads 300°C, a platinum-resistance thermometer in the same place and at the same time will read 291°C! As an advanced level physics textbook points out, "No one of them is any more 'true' than the other, and our choice of which to adopt is arbitrary, though it may be decided by convenience."

Most people who take for granted that temperature can be measured very accurately, objectively and with ease are unaware of these complications, which are similar to complications which arise in the measurement of intelligence. Different types of test are adapted for different purposes and for different populations, and they do not always give identical results, any more than do different types of thermometer. This does not prove that intelligence cannot be measured, or that our measurement is not objective and scientific; it simply means that it is beset with the same problems and difficulties as the measurement of heat or any other physical quality.

A DISCIPLINE IN ITS INFANCY

Can we go beyond the operational definition of intelligence, or the simple verbal statement that it is "general cognitive ability"? If at present we have no widely agreed general definition, this is by no means a death blow to the concept itself; the same is true of practically all scientific concepts. For instance, there are three different theories and definitions of gravitation. The first is Newton's original action-at-a-distance theory. The second is Einstein's "field" theory. The third, based on quantum theory, treats the interaction of bodies as analogous to the other fundamental forces of nature—the strong nuclear, the weak nuclear and the electromagnetic force—and explains gravity in terms of an elementary (but possibly imaginary) particle, the graviton.

The fact that physicists have no final, universally agreed theory of gravitation has not meant that attempts to measure the force of gravity have not been scientific and successful in practical terms. Universally agreed definitions come at the end, not at the beginning, of scientific research; even after 300 years of work in the field of gravitation, by some of the most brilliant scientists of all time, a simple answer still eludes us. Should we expect more of scientists in psychology, which is possibly a much more difficult field, and in a much shorter period of time? We shall come back to the definition of intelligence, and try to elaborate it, later on. Let us merely conclude for the present that the abstract quality "g" can be identified and that few people would dispute that it can be measured reliably and validly by means of traditional intelligence tests. We must now turn to the question of whether "g" can be identified with intelligence as it is popularly understood.

4

INTELLIGENCE AND ACHIEVEMENT

The question of whether or not "g" corresponds to popular notions of intelligence is complicated by the fact that popular notions of intelligence are inconsistent and sometimes contradictory. What the man in the street has to say about intelligence boils down to two rather different things.

LAY IDEAS OF INTELLIGENCE

In the first place, the layman identifies intelligence with problem-solving ability, with cleverness, with thinking; that is, with precisely the type of cognitive behaviour which Cicero labelled *intelligentia*. This popular notion of intelligence is very similar to the concept of fluid intelligence: an ability to solve problems that can be applied to any kind of situation.

The other popular definition of intelligence is acquired knowledge. A person who is learned in some respect—who has an academic degree or a diploma, or has acquired in some other way a reputation for being knowledgeable—is considered "intelligent", regardless of whether or not he is adept at problem-solving. This corresponds quite closely to the concept of crystallised ability we have come across before.

ABILITIES OVERLAP

Tests of mental ability usually measure both fluid and crystallised ability. Thus the Raven test, introduced by John Raven and widely used by the armed forces, in school selection and for other purposes, consists of two parts: one, called the Progressive Matrices test, is a test of fluid ability, while the other, a vocabulary test, is a test of crystallised ability.

We might expect these two tests to be uncorrelated, because acquired knowledge and problem-solving ability seem to be quite different things. Yet the tests correlate quite highly, as indeed do all tests of fluid and crystallised ability. The reason is very simple. If you have a high degree of fluid ability, then, other things being equal, you are likely to acquire a greater degree of knowledge than someone with less fluid ability. You will tend to acquire a better vocabulary. This is partly because you are more likely to be interested in a wide range of information, will read more newspapers, journals and books, and will listen to more lectures and programmes of cultural or scientific interest. It is also, and equally important, because your intelligence will help you to understand and remember, in an ordered sequence, the items of information, including vocabulary, you come across. You will, in other words, develop more crystallised intelligence.

So far there seems good reason to equate "g" with intelligence. But further proof is needed. The man in the street would expect an intelligent child to do better at school than a dull one; an intelligent adolescent to do better at university than a dull one, or indeed an intelligent one to proceed to university, where a dull one would fail. He would expect an intelligent person to go into a higher-level occupation like medicine, the law, or science, and a dull person to go into an unskilled or semi-skilled job. Before "g" can be equated with the layman's idea of intelligence, IQ tests must, at the very least, demonstrate that these predictions are fulfilled. If IQ tests correlate with other measures which can be taken to indicate intelligence, such as educational level and social standing, they are said to have *external validity*. What are the facts?

IQ AND SCHOOL SUCCESS

There is no doubt that a reasonably close relationship exists between high IQ and success at school, if success is measured by both marks gained and duration of schooling. Pupils with high IQs tend to gain high marks and to stay longer at school; those with poor IQs tend to do poorly in their class work and to drop out earlier. These relationships have been observed unfailingly over many years and in many countries. Correlations are highest for the most academic subjects, like Latin, and lowest for the least academic, like gymnastics. They may even disappear for quite unacademic subjects, although even for subjects such as sewing and cooking, small correlations usually persist.

Distorting factors
The size of the correlation observed between IQ and scholastic success varies very much from one study to another, for a variety of reasons, including selection procedure, teaching policies and motivation.

Different principles of selection are applied in different schools, in different countries, and for different subjects. The greater the selection, the more uniform the IQ level of a given class is likely to be; and, by the very nature of statistics, the smaller the range of IQs, the lower the correlations with success will be. In Britain, the typical non-selective, unstreamed comprehensive school would be expected to produce larger correlations than the typical highly selective, streamed "public" (fee-paying) school. By and large these expectations are borne out.

Correlations may be reduced from the expected level by certain policies adopted by a school or by individual teachers. In some comprehensive schools teachers pay much more attention to dull children than to bright ones, attempting to bring them up to the average level of the class. Brighter children may be prevented from going ahead too fast, which reduces the level of achievement of the class as a whole, and with it the correlation between IQ and achievement. Teachers have even been known to give identical marks to all children, on some ideological principle which rewards effort rather than achievement, thereby making correlations disappear altogether.

Another distorting factor is motivation. In a mixed-ability class the bright ones may be bored because the teacher goes over the same material again and again for the sake of the duller ones, and the dull ones because they can't understand the material however many times it is repeated. This often leads bright pupils into truancy, cheekiness and other misdemeanours, and distracts them from academic work.

It will be clear that intelligence is necessary for high-level school and academic work but not sufficient on its own. Other factors also play a part. One of these is persistence and hard work: achievement requires application as well as sheer ability. Personality is another. Introverts tend to do better at academic work than extroverts, and people showing emotional instability tend to do poorly. All this leads to a somewhat asymmetrical relationship between intelligence and achievement. In other words, high achievers are practically always very bright, and low achievers tend to be dull. Some low achievers, however, are found to be bright but lacking in persistence and application, or neurotic, or extroverted, or failures for some other reason unconnected with intelligence.

The Eleven Plus

Curiously enough, lack of a perfect correlation between intelligence and scholastic success is sometimes advanced as a criticism of intelligence tests. In England, selection for different types of secondary education used to be carried out by the Eleven Plus examination but because prediction was less than perfect, the method was severely criticised and finally abandoned. Some of the opprobrium attaching to intelligence tests today stems from this experience, yet it is completely misplaced. In the first place, the examination itself was not an intelligence test; it consisted of three papers, one in English, one in mathematics, and one a verbal reasoning test which could be considered a test of crystallised ability very much dependent on acquired knowledge. There was no test of fluid ability included in the Eleven Plus examination at all. In any case, even at its best a test of intelligence only measures one of the variables which determine academic success—admittedly an important variable, possibly the most important single variable, but nevertheless only one of several. It is quite unrealistic to expect perfect predictions under these circumstances. Indeed, if predictions had been perfect, they would have disproved the very theory on which they were based, because they would have equated a latent trait (intelligence) with an overt trait (achievement).

At this stage readers may well ask why in the selection process the test used was one of crystallised ability, heavily dependent on acquired knowledge and therefore to some extent culturally biased. The answer of course is that the educationalists who put the Eleven Plus examination together, and who construct similar tests in other countries, are not concerned with pure, scientific measurement, but rather with prediction. They prefer such a test to a pure test of fluid ability because it draws on a mixture of pure intelligence and acquired knowledge which gives a better prediction of academic achievement. Most psychologists would probably say that a better method of prediction would be to administer pure tests of fluid ability and tests of academic achievement separately, then to combine the scores. The so-called IQ tests used in education, industry and elsewhere are not, truly speaking, tests of fluid ability and therefore only deserve the title "intelligence test" by courtesy. Criticisms that these tests are culture-bound are often justified but would not apply to proper tests of fluid ability.

IQ AND ACADEMIC SUCCESS

What has been said of children's academic achievement applies equally well to the academic achievement of students. Here, too, there is

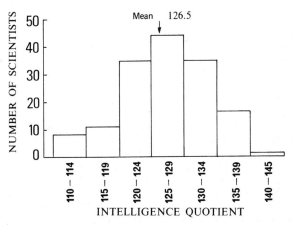

Fig. 5. Wechsler test IQs of 148 faculty members in various science disciplines at the University of Cambridge. (Adapted from Gibson and Light, 1967)

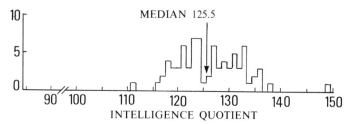

Fig. 6. Wechsler test IQs of 80 medical students. (Adapted from Kole and Matarazzo, 1965)

a correlation between intelligence and success, but of course it is not very high simply because the range of intelligence of students accepted for university work is quite limited: hardly any will have IQs below 110 or · 115. Correlations have been worked out for literally thousands of students, but they vary from one university to another, depending on the range of ability admitted. Again we find an uneven relationship between IQ and achievement; intelligence is necessary but not sufficient for high academic success. Figures 5 and 6 show the distribution of IQ on the Wechsler test (which consists of ten sub-tests covering many different types of problem) of 148 faculty members in various science disciplines at the University of Cambridge, and of 80 medical students. These are pretty typical of distributions observed elsewhere among university staff and students.

IQ AND OCCUPATION

Turning to occupations, we would expect people in middle-class jobs to have higher IQs on average than people in skilled working-class jobs, and the latter higher IQs than people in semi-skilled working-class ones. This expectation is indeed borne out; table 1 shows the mean IQs of a number of different occupations in the US. Similar figures were obtained during the First World War from soldiers entering the armed forces from various occupations. The scores given in Figure 7 were obtained directly from a test called the Army Alpha and are not conventional IQs. But they show similar differences between various classes of occupation, from the middle-class engineer down to unskilled labourer. Upper middle-class professionals such as university professors and medical consultants would go at the top of this league, with IQs in the 135–140 range.

Army figures show a similar distinction between enlisted men, corporals, sergeants and officers. Figure 8 gives the Army Alpha scores of various groups in the First World War. This was the first occasion on which intelligence tests were used—with great success—to select officers and non-commissioned officers. The British army was forced to adopt a

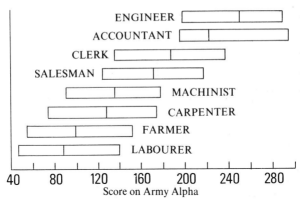

Fig. 7. Scores on the Army Alpha test of First World War soldiers entering service from various occupations. (Adapted from Yerkes, 1921)

Table 1. Average IQs of members of different occupations in the US. (Adapted from Harrel and Harrel, 1945)

	MEAN:	
ACCOUNTANT	128	
LAWYER	128	
AUDITOR	125	
REPORTER	124	MIDDLE-CLASS
CHIEF CLERK	124	OCCUPATIONS
TEACHER	122	
DRAUGHTSMAN	122	
PHARMACIST	120	
BOOK-KEEPER	120	

TOOLMAKER	112	
MACHINIST	110	
FOREMAN	110	
AIRPLANE MECHANIC	109	SKILLED
ELECTRICIAN	109	WORKING-CLASS
LATHE OPERATOR	108	OCCUPATIONS
SHEET METAL WORKER	108	
MECHANIC	106	
RIVETER	104	

PAINTER, GENERAL	98	SEMI-SKILLED
COOK AND BAKER	97	WORKING-CLASS
TRUCK DRIVER	96	OCCUPATIONS
LABOURER	96	
BARBER	95	
LUMBERJACK	95	
FARMHAND	91	
MINER	91	
TEAMSTER	88	

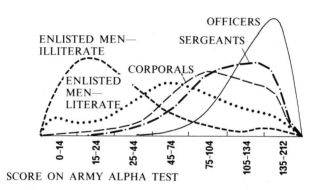

Fig. 8. Intelligence test scores of various US Army groups during First World War. (Adapted from Yoakum and Yerkes, 1920)

similar method some time after Britain entered the Second World War, because other methods had failed to produce successful officer candidates, and it continues to use it to this day. Many other countries have followed suit.

IQ AND STATUS

Would it be true to say, in general, that the prestige, the income and the intellectual requirements attached to an occupation are highly correlated? The answer needs to be Yes for us to say that IQ tests really measure what the man in the street regards as intelligence. Large-scale investigations show that this is, indeed, the case.

Firstly, the Barr Scale of Occupations was drawn up by a number of psychologists who rated 120 representative occupations for the grade of intelligence ordinary success in each required. Secondly, there are the results of a large public opinion poll by the National Opinion Research Centre (NORC) in which numerous occupations were rated for prestige. Lastly, we have ratings of socio-economic status (SES), as assigned officially in the 1960 US census of population; hundreds of occupations are listed on the basis of their average income and educational level.

The prestige rating of an occupation and its intellectual requirements as determined by NORC and Barr respectively correlate 0.91; prestige and income correlate 0.90; intellectual requirements and income correlate 0.81. There is thus a close relation between the intelligence needed in an occupation, the social prestige attached to it, and the income and education of the people in it. If we regard income and prestige as having social importance, then it is obvious that intelligence precedes occupational choice, and is thus clearly implicated in the other two factors.

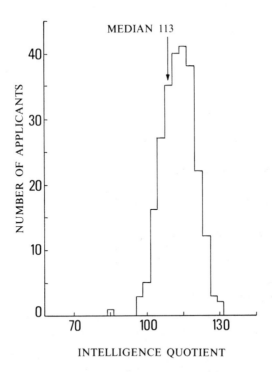

Fig. 9. Wechsler IQs of 243 police and fire service applicants. (Adapted from Matarazzo, 1964)

Success at a given job

These are differences in intelligence between occupations. Are there similar differences within occupations—between those who succeed particularly well, and those who do less well in a particular occupation? The answer is that the correlations, although positive on the whole, are not very large. There are two reasons for this. In the first place, the range of intelligence within a given occupation is relatively small. Figure 9 illustrates this, using as an example the Wechsler IQ scores of 243 applicants to the police and fire service. The range is only about 30 IQ points, compared with a range over three times as great in the population at large. This restriction inevitably lowers any correlation between IQ and success in an occupation.

The second point is, of course, that many outside forces make it difficult for people to distinguish themselves in a given occupation. Trade union rules may force people to work to a lower standard than they would have chosen for themselves in order not to show up the less able or the less willing. And in many occupations it is difficult to establish a degree of excellence: who is to tell which of several medical practitioners is in fact the best doctor? The criterion of professional distinction is difficult to establish and not always reliable.

The general congruence between IQ and income does break down occasionally, for obvious reasons. There are groups of people whose earnings bear no relation to their intelligence—actors, tennis players, prostitutes, TV personalities, royalty, disc jockeys, gigolos and golfers, for instance. But the numbers involved are quite small and do not invalidate the overall conclusion. Luck, nepotism and similar factors also make the correlation less than perfect, as does the impact of personality and other influences.

INTELLIGENCE AND "G" EQUATED

This is a brief summary of literally hundreds of studies which have investigated the relationship between IQ, educational success and general success at living. All of them demonstrate a positive relationship varying in strength according to the factors mentioned. There seems little doubt that IQ tests do measure what the man in the street would identify and recognise as intelligence. This would seem to justify us in identifying "g" and IQ with intelligence, and in using the terms "g" and intelligence interchangeably, which we shall do for the rest of this book.

5
SEX,
AGE AND
INTELLIGENCE

Binet's work, and the concept of the IQ, were of course based on the notion that intelligence develops with age, increasing up to late adolescence and possibly a little later. This increase in intelligence with age (not, as some critics have suggested, differences between middle-class and working-class groups) was the first and main criterion by means of which test items were judged.

Though formally independent of the internal and external criteria discussed in previous chapters, this criterion fortunately agrees with the conclusions derived from both these other sources: an IQ test which does well on the internal criterion of correlation with other tests, and on the external criterion of producing large differences between those who are successful and those who are not successful in educational and academic pursuits, also usually shows high correlations with increasing age. It is agreements of this kind between formally independent criteria that make the paradigm of modern intelligence testing so strong.

THE AGE FACTOR

Figure 10 shows an interesting test which illustrates the increase of ability with age. The ten figures which have to be copied by the child all seem so easy that one might think there would be no difference between them. In fact, there is a rigid age sequence, with children becoming able to copy the more difficult ones only as they advance in age. It is possible (though very difficult) to teach a child to do a test item in advance of his mental age, but once he stops practising he soon relapses and falls back into his age group. Much the same is true of the various items in the Binet test, and also of the rather novel type of item the Swiss educationalist Jean Piaget has been working with in elaborating his own theory of mental development.

Rise and fall of test scores

Given that intelligence advances with age up to young adulthood, what happens as a person grows older? Does performance on all types of

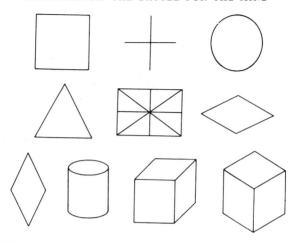

Fig. 10. Gesell figure-copying test

test decrease equally with age, or more quickly on some than on others? Figure 11 shows the development and decline of scores on the Wechsler test with age. Though these scores are measurements of intelligence, they are not conventional IQ scores, which have a mean of 100. The

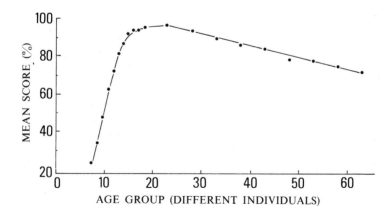

Fig. 11. Wechsler scores, showing growth and decline with age. (Adapted from Matarazzo, 1972)

progression is very much as anticipated. Performance improves up to between 16 and 20 or thereabouts. Then it declines in fairly regular fashion. In looking at this figure, it is important to realise that the curve represents the averaged results of a number of different tests. The Wechsler test is made up of ten quite different sub-tests, some verbal, some non-verbal, some pencil-and-paper, and some using apparatus. Performance on the different tests declines at different rates: on a crystallised ability test such as the vocabulary test it shows little if any decline, and it shows the most on a test of fluid ability such as the Block Design test, in which the subject is given a number of blocks, with a different pattern on each face, and is asked to copy a particular pattern. There are marked differences in the rate of decline of different types of intelligence, and it is important to bear this in mind in assessing a person's chances of succeeding at an academic or intellectual job at any given period of his life.

The growth curves of different abilities are different too, as Thurstone was the first to show. His estimates are shown in Figure 12. It will be seen that perceptual speed grows most quickly, word fluency more slowly. The differences are noticeable but not overwhelming—all abilities follow a rather similar growth, with minor variations. Raymond Cattell, another of the giants in the field of IQ testing, has suggested a general difference in the growth of crystallised and fluid ability with age. With crystallised ability, as Figure 13 illustrates, the terminal level—maximum development—is reached later by the more able. Growth in fluid ability, as Figure 14 shows, ceases at around the same time for the able, the average and the less able.

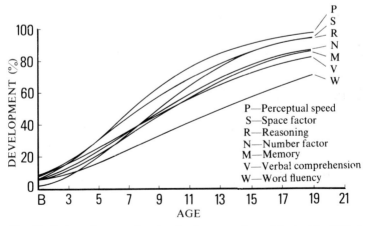

Fig. 12. Estimated curves for the development of special mental abilities. (Adapted from Thurstone, 1955)

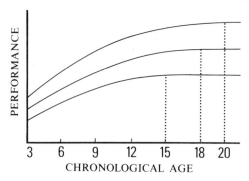

Fig. 13. Growth of crystallised ability with age; terminal level is reached later by the more able. (Adapted from Cattell, 1971)

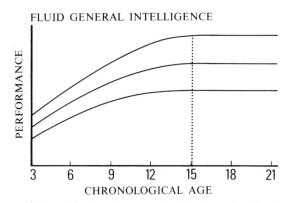

Fig. 14. Growth of fluid ability with age; terminal level is reached simultaneously by bright and dull. (Adapted from Cattell, 1971)

THE SEX FACTOR

While age differences are large and important, sex differences are relatively small. On practically all the IQ tests now in wide use, men and women have equal average scores. This is sometimes attributed to some kind of chicanery on the part of psychologists. They are said to have selected items in such a way that equal scores are achieved regardless of whether there might or might not be genuine differences between the sexes. This accusation is false. Tests such as the Matrices tests, the

Dominoes and many others were constructed quite irrespective of sex, and were found to give equal scores to boys and girls, to men and women. Given that unselected items give the sexes equal IQ scores, it was only reasonable for other test designers to avoid bias in favour of one or the other sex by making certain that their tests followed the same line.

If tests sometimes favour one sex, it is simply because the sexes do differ slightly in primary abilities, and if a test contains too many items relevant to an ability in which one sex is generally superior, then the total score may be affected. This imbalance can be avoided by suitable selection, but not all test constructors are as careful, dependable and knowledgeable as they might be; it is useful to have a final check to make sure that sex bias has in fact been avoided in tests which include items covering primary abilities. This precaution has been taken with the Wechsler test, for instance. It is worth stressing that tests like the Matrices and Dominoes are measures of pure "g" and do not therefore distinguish between the sexes. Tests like the Wechsler, which measure primary abilities, need to be carefully balanced to avoid sex bias.

Men are better at spatial tasks

Generally men exceed women in visuo-spatial ability, that is the ability to organise and manipulate visual inputs in their spatial context. Men are better than women at perceiving patterns as a whole, and consequently at such practical skills as map-reading and mechanics. Animals such as chimpanzees and rats show the same sex-related differences in visuo-spatial ability, which does not seem to be affected much by cultural factors. This may be related to evolution: the male animal needed to maintain accurate spatial orientation during his foraging, and to detect spatial relationships despite distortions and camouflage. There is evidence that the ability is not only genetic but also to some extent sex-linked, and that it develops under the partial control of the sex hormones.

Women are better at verbal tasks

If men are superior in visuo-spatial ability, women show almost the same degree of superiority in verbal ability. Girls learn to talk earlier than boys, and they articulate better and possess a more extensive vocabulary at all ages. They write and spell better, their grammar is better, and they construct sentences better. These differences can be observed as early as six months! In other species, particularly those where emotions are indicated by vocalisations, females also show pronounced superiority.

But though females are superior in language usage, or verbal fluency, they are not superior in verbal reasoning, meaning the use of intelligence in problems which are presented verbally. When comprehension and reasoning are taken into account, boys are slightly superior to girls. Females are also better at learning by rote. They seem able to memorise for short periods a number of unrelated and personally irrelevant facts, while men are capable of comparable feats only if the material is personally relevant and/or coherent. (This is probably exactly the opposite to what most people would have thought intuitively.) This ability, too, appears to be genetic.

Convergence and divergence

Boys and girls also differ to some extent in what may be called *cognitive style*. Test items can be divided into two categories—*convergent* and *divergent*. Examples of convergent items are those given in Chapter 2. The relations between the components point to one single correct solution; they converge on this solution. A divergent item, on the other hand, has no single correct solution; it has an infinite number of correct solutions, and the score is the number discovered by the person tested. "How many uses can you think of for a blanket?" is a divergent question. Tests of this kind are sometimes called "creativity" tests, on the assumption (for which there is some slight evidence) that they measure originality as well as intelligence.

Boys seem to have a more divergent cognitive style, a difference which can already be seen in the play of children under school age. It is hard to know whether to attribute the difference to originality or to girls' greater reluctance to make nonsensical suggestions. There has been some resistance on the part of psychologists to investigating sex differences, for fear, no doubt, of upsetting the egalitarian applecart. This is unfortunate: recognising possible differences between the sexes does not entail assigning superior status to one or other. A rational view of equality does not demand identity, and for practical purposes it is important to know in what ways women do better than men, or men than women. Whatever the findings, they could hardly affect the issue of overall equality of mental ability between the sexes.

Geniuses and defectives

While men and women average pretty much the same IQ score, men have always shown more variability in intelligence, as in many other physical and mental traits. In other words, there are more males than females with very high IQs and very low IQs. This accords with the common observation that far more geniuses in science, the arts and other pursuits, and far more mental defectives, are men than women. Figure 15 illustrates the difference in diagram form.

It is of course possible to think of environmental reasons why this should be so. The pressure of child-bearing and traditionally feminine tasks, as well as male opposition, may have made it extremely hard for women to devote all their energies to scientific or artistic pursuits and therefore to achieve the highest distinction. The pressures of earning a living may have led to quicker recognition of mentally defective males. Mentally defective women, on the other hand, may have been able, if at all attractive, to escape institutionalisation by marrying. It would be difficult to prove or disprove these possibilities. However, there is an

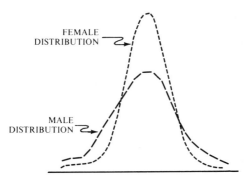

Fig. 15. Distribution of male and female IQs, showing greater variability of males. Differences are exaggerated for clarity. (Adapted from Lehrke, 1979)

alternative explanation for which there is some evidence. It is based on the genetic concept of sex-linkage.

How sex-linkage works

Sexual differentiation in higher animals depends on the sex chromosome complement—two X chromosomes for females, and an X and a Y for males. The X chromosome in man is of medium size, containing about 5 or 6 per cent of the genetic material and carrying about the same proportion of genetic information including genes known to affect every major body system. The Y chromosome, on the other hand, is one of the smallest, and, as far as is known, carries only genetic instructions for maleness. While all the other chromosomes operate in pairs, in the case of the sex chromosomes, this pairing occurs only in females—in males the Y chromosome does not pair up with the X chromosome. As a result, in males whatever genes are contained in the X chromosome will find expression without any interference from the Y chromosome, while in females the second X chromosome will reduce the impact of whatever genetic information is contained in the first. This makes males genetically more likely to show some excess at either end of the scale, and accounts for their greater variability, as Robert Lehrke has pointed out. We are now in a position to suggest that male–female differences in IQ variability may have a genetic basis in sex-linkage.

IQ in the family: the power of the X chromosome

This hypothesis can be tested directly. What one would expect, if there are major genes relating to intelligence on the X chromosome, is that the correlations of test scores for mother–daughter, father–daughter and mother–son would be quite similar, because in each case the parent and child have one X chromosome in common. However, correlations between fathers and sons should be lower since they have no X chromosome in common, and brother–sister correlations should be intermediate since they have an X chromosome in common half the time.

Bayley (1966) has provided relevant data. She found a mother–daughter correlation of 0.68, a father–daughter correlation of 0.66, a mother–son correlation of 0.61, and, as expected, a very much lower father–son correlation of 0.44. Brother–sister correlations of 0.55 were found. In other words, the order of size of these correlations is precisely what would be expected on the basis of an X-linked trait. A good deal more information is available on this and similar points, all of which supports the hypothesis. Lehrke discusses both the theory and the evidence at length.

Male variability: what it means in numbers

The actual difference in variability between the sexes is relatively slight, which might lead the reader to suppose that the issue is not an important one. However, because of the mathematical properties of the normal curve of distribution—the bell-shaped curve used to represent IQ distribution is an example of a "normal curve"—differences in variability have a greater effect the further away they are from the mean; in other words, effects would be much more marked with very high and very low ranges of IQ. Let us assume that the variability of males is 13 per cent greater than that of females, a difference slightly lower than the one actually observed in the 1947 Scottish Survey which was on a very large scale. On this basis we would expect 37 per cent more males than females with IQs below 68 or above 132. In the really high-IQ range, the difference would be far greater even than that, but no precise figure will be offered here because it is doubtful whether the curve of distribution of IQ remains normal above the level of 130 or thereabouts.

However that might be, it seems likely that there is some degree of sex-linkage in intelligence, and that this accounts for the greater variability of males as far as intelligence is concerned. Such a finding has no bearing on the question of who are the more intelligent, men or women. Dr Samuel Johnson, when asked this question, replied: "Which man? Which woman?" It is difficult to think of a better conclusion to this chapter. Statistical truths about averages make no predictions about individuals and should not be interpreted as doing so.

6
IS
INTELLIGENCE
INHERITED ?

The form of this question, although not unusual, is misleading and ultimately meaningless. In one sense of the term, intelligence, or the possibility of solving problems successfully, is obviously inherited; it is one of many things which distinguish human beings from earthworms and stones, and as such the basis of this ability is clearly genetic.

What is really meant by the question, of course, is: "Are *differences* in intellectual ability between human beings determined genetically?" In this form the question is still misconceived, because it suggests that such differences are caused either by genetic factors, or by environmental factors; in reality it is exceedingly unlikely that genetic or environmental factors by themselves could be responsible for the differences observed in intelligence, and the question really should be: "To what extent are the observed differences in human ability due to environmental factors, to what extent are they due to genetic factors, and to what degree is there interaction between the two?"

A COMPLEX ISSUE

Even in this form it is grossly oversimplified. There are different *types* of genetic factor, and we should be concerned with clarifying the degree to which these may play a part in the determination of individual differences in intelligence. Thus we may ask whether high intelligence is genetically dominant over low intelligence, or we may ask whether there is evidence of assortative mating as far as intelligence is concerned—that is, do bright men marry bright women? (Genetic dominance and assortative mating are two important genetic factors that will be discussed later on.) Similarly, there are different forms of interaction between genes and environment, and last but not least, environmental differences can be classified in various ways.

Thus a geneticist differentiates between within-family environmental factors (those which distinguish one member of a family from another) and between-family environmental factors (which distinguish one family from another). The latter include socio-economic status, the number of books in the home, educational pressures in the family and so on. Even within the same home, not all the children have the same environment. They may attend different schools. One child may be ill at a crucial period of his school life, while another is healthy. One may find a suitable boyfriend or girlfriend, while the other may find an unsuitable one. These are within-family environmental differences.

Certain environmental factors go beyond these simple differences. Some environmental factors may be intra-uterine: the foetus may be infected by syphilis even before birth, may be affected by drug-taking on the part of the mother, or may, in the case of identical twins, suffer from the "transfusional syndrome", in which one twin causes toxic effects in the other as a result of difficulties in the blood supply. As to post-natal factors, these include physical influences (malnutrition, sensory deprivation—keeping a child locked up in a cellar is an extreme example—or illnesses affecting the central nervous system) and non-physical ones, such as differences in education, motivation and so on. When we talk about "heredity" or "environment", we should always be careful to specify which of these many different factors we are referring to.

Types of interaction

Even a concept like interaction between heredity and environment is more complex than it might appear at first sight, for there are different types of interaction. In the first instance, we have the simple *additive relationship* between heredity and environment; when we say that differences in variance in intelligence are attributable 80 per cent to heredity and 20 per cent to environment, we imply that both contribute to produce the results observed. In addition, there is the possibility of *statistical interaction*—the possibility that different genes may respond differently to the same environmental effect. There is some evidence to show, for instance, that glutamic acid increases the IQs of dull children, but not of average or bright ones; this would be an instance of statistical interaction. Then we have what are sometimes called *correlated environments*. An example is a child with genes for high intelligence who is also reared in a home offering superior opportunities for intellectual development; this is a rather different type of interaction again.

Even this sketchy look at the complexities of the problem is enough to show that discussions about heredity and environment which are not informed by a deep understanding of the issues, or a knowledge of the statistical complexities involved in analysing out these various factors from empirical observations, must by the nature of things be at best irrelevant and at worst misleading. Unless we specify precisely, and preferably in quantitative terms, just what it is that we are talking about, discussions about nature and nurture almost necessarily deteriorate into ideological statements and political dogma. Both aspects of behavioural genetics—the genetic theory and the statistical analyses—are complex.

All that this book can do is to discuss the kinds of evidence which have been considered to have a bearing on the issues. (Readers wanting a proper introduction to the field, with references, are referred to my book, *The Structure and Measurement of Intelligence*.)

METHODS OF INVESTIGATION

It is important to emphasise two points. The first is that there is not just one method of investigation but many, and these methods complement each other in the sense that they throw light on different aspects of the general problem. It is quite wrong, for instance, to think that twin studies are the only method which has been used, or can be used, in this connection; twin studies are important and often convenient, but other methods are equally, or even more, valuable. The second point is that quantitative estimates of heritability and other aspects of the nature–nurture problem can be derived from many different sources and methods of investigation; they become credible and acceptable because, *regardless of the method used, they arrive at pretty similar values*. Like all scientific measurements, these estimates are subject to measurement errors which are larger than we would like, although they are still smaller than measurement errors in some of the so-called "hard" sciences. (Psychology is regarded as a "soft" science because not all the concepts in which it deals have yet been satisfactorily quantified.) Our estimates of heritability are almost certainly closer to the true values than are estimates of stellar distances in astronomy, for example. But margins of error always exist in scientific measurements, and errors in the field of intelligence are still too large for comfort, and need to be reduced.

The logic of twin studies

To take twin studies first, we must begin by outlining the logic of twin research. Twins are divided into monozygotic (MZ) or identical twins and dizygotic (DZ) or fraternal twins on the basis of similarity or dissimilarity of physical characteristics which are known to be very highly genetically determined, such as facial appearance, fingerprints, or blood group antigens. As they are the product of the same ovum which has split in two, MZs are genetically identical, whereas DZs are no more alike genetically than other brothers or sisters. Subjects are measured on the trait under investigation, and the extent to which MZ twins are found to resemble each other *more* than DZ twins is taken as an indication of the genetic contribution.

Consider an early investigation carried out in England by Herrman and Hogben. They studied MZ twins, DZ twins of the same sex, DZ twins of different sex, and siblings (ordinary brothers or sisters). They ascertained the average (also called "mean") difference in IQ between twins or siblings. For the 65 pairs of MZ twins, the mean IQ difference was 9.2. For the 96 DZ twins of the same sex, it was 17.7, and for the 138 pairs of DZ twins of different sex, it was 17.9. For siblings, the mean IQ difference was 16.8.

These results are typical of much later work, and they are quite clear-cut. First of all, there is no difference between DZ twins of the same sex and DZs of different sex, which suggests that genes and environment operate on both sexes in the same way. Second, DZ twins are no more alike than ordinary siblings, which indicates that twins are not treated differently from ordinary brothers and sisters in any ways that would affect intelligence. Third and most important, MZ twins are much more alike than DZ twins or siblings, the average difference for MZs being only about half that for the others. Since MZ pairs are genetically identical, and DZ pairs are not, it is plausible to ascribe their greater IQ resemblance to genetic influences.

Being treated alike is of no consequence

The assumption in all this is, of course, that DZs share relevant environmental influences to the same extent as do MZ twins; if this assumption does not hold, the increased resemblance of MZ twins may simply reflect their greater environmental similarity. There is some evidence to suggest that MZ twins are treated more alike than DZ twins, in the sense of dressing alike, playing together, sharing the same teacher, sleeping in the same room, and because of conscious attempts by parents to treat them alike. However, the important question is whether or not such differences in treatment are important determinants of intellectual ability: if they do not influence IQ, they are irrelevant. In a large-scale study based on over 2,000 pairs of twins, Loehlin and Nichols showed that these influences had absolutely no effect: those twins who were treated more alike were not more alike in intellectual ability.

Identical twins—an intriguing blend of solidarity (spectacles, hairstyle, playing together) and individuality (different clothes).

This was not surprising if we recall Herrman and Hogben's finding that neither sharing the same gender, nor being a DZ twin as opposed to an ordinary sibling, influenced similarity of IQ in any way. If being treated as a boy rather than a girl does not affect similarity in IQ, it is not surprising that dressing similarly or playing together more had no effect. There is no evidence at all to suggest that MZs being treated differently from DZs affected their cognitive development in any way.

It would be tedious to list the many studies which have followed the research of Herrman and Hogben; the quantitative results are always very similar. In terms of correlations, that for MZ twins in the Herrman and Hogben study was 0.84, and for DZs 0.47; the average for a large number of studies was 0.87 for MZs and 0.53 for DZs. When a statistical analysis is made of these figures, it suggests that of the total variation in IQ something like 70 per cent is due to genetic causes, at the most 20 per cent (actually 19 per cent) to characteristics of the family environment, and about 10 per cent to the particular way the individual is treated in the family. These figures require some correction because in the usual formulae errors of measurement are counted as part of the environmental factors, and we should really only concern ourselves with the part of the total score which is error-free; correction would then increase the genetic part of the total variation to something like 80 per cent.

MZs reared apart
Another way of using twins is to look at MZs brought up separately. Three major studies have been reported, showing correlations of about 0.77 (the results of all three studies are in very close agreement). A fourth study has been reported by Sir Cyril Burt, also giving the figure 0.77, but doubt has been thrown on the genuineness of his data, and they will therefore be omitted from consideration. Obviously, though, including his data would have made no difference to the figure arrived at.

Taken at face value these tests suggest a heritability of about 77 per cent, but this is almost certainly an overestimate because the environments in which the twins were brought up were not usually random: their environments were more similar than they would have been if the twins had been unconnected. On any reasonable assumption, correcting for this would bring the correlation down to something like 68 per cent, which in turn needs correcting (upwards) for unreliability. Two of the three studies of identical twins brought up in isolation used individual tests rather than the group tests we have been discussing so far (in other words, each subject had a test administered to him personally by a tester), and these gave correlations between MZ twins of 0.67 and 0.68.

Siblings reared apart
There are a few studies of siblings reared apart; these give a slightly lower estimate of heritability than studies of twins, but not strikingly so. More numerous are studies of unrelated individuals reared together (foster-siblings). Their resemblance will be a pure reflection of shared

environment, provided there is no selective placement. (Selective placement is the practice of matching the background of adopting parents to that of the child's biological mother.) Where there is selective placement, the results would overestimate environmental effects, and underestimate genetic influences.

Seven such studies have been reported, and they give a median correlation of 0.23. (Whereas the average or mean of a series of figures is the total divided by the number in the series, the median is the middle figure: the average of 4, 8 and 9 is 7, but the median is 8.) This figure of 0.23 is a direct estimate of environmental variation between families and is only trivially different from the 19 per cent suggested by the MZ and DZ twin data mentioned earlier; the fact that it is slightly higher is probably due to selective placement. Broad agreement between the various lines of evidence is beginning to emerge.

PARENTS AND PARENTING

Fostering has a minor influence

Comparisons between foster parents and their children also allow us to estimate the sources of IQ differences in a quantitative manner. Many such studies, involving well over 1,000 pairs of children in all, have been published. In the absence of selective placement, the correlations obtained are a direct estimate of the effects of home environment. The IQ of adopted children produced a median correlation of 0.17 with the IQ of foster fathers and 0.21 with that of foster mothers, the overall median value being 0.19. This agrees quite well with estimates of the contribution of family factors (between-family environment) obtained from the other lines of evidence mentioned.

Natural parents versus foster parents

If the correlation between a child and his or her *natural* parent is compared with the correlation between a child and his *foster* parent, we can get an idea of the influence of genetic factors. Whereas, as we have seen, the correlation between foster parent and child is 0.19, a median correlation of 0.50 was found in 12 studies of natural parents and their children. Estimating heritability from these data gives us a figure of 62 per cent, which is somewhat lower than the true figure because of selective placement, but even so quite close to the figure of 68 per cent heritability yielded by the studies on MZs brought up in isolation. It will be remembered that all these figures require upgrading because of the unreliability of the tests; I have not bothered to carry out the necessary calculations but have instead compared the values found in these various investigations with the "raw" or unconverted estimate derived from the twin data—68 per cent. The corrected values would always be 10 per cent to 12 per cent higher.

The most direct evidence of the genetic component in parent–child resemblance comes from studies of natural parents and their children given up for adoption shortly after birth. Only three such studies have been reported, and the results suggest a heritability of 64 per cent. Again, agreement between the various lines of evidence is very good. One of these studies, the 1949 study by Skodak and Skeels, tested the children at various ages, so that correlations could be made between the natural mother's IQ and the IQ of the child at two, four, seven and 14 years of age. The pattern is very clear: heritability is nil when the children are only two years old, but rises steadily to 80 per cent by the time they are 14. Finding that a delay *increases* the resemblance in IQ is strongly suggestive that the cause of this resemblance is genetic.

Also reported were correlations, at various ages, between the child's IQ and the educational level of the foster parents. Once again the pattern is very clear: at no time do adopted children and foster parents correlate more than 0.1, and *adopted children do not grow to resemble their adoptive parents*. This is in marked contrast to the fact that children certainly do grow to resemble their natural parents, even when they do not live with them and have been separated from them since shortly after birth. It would be difficult, in the face of these findings, to deny the presence of a strong genetic component in parent–child resemblance.

KINSHIP EVIDENCE

Another source of evidence is IQ comparisons of blood relatives— what are called kinship correlations. The degree of consanguinity between two relatives should determine the similarity in their IQs. For instance, two brothers should resemble each other more than first cousins. A good many kinship studies have been carried out, and by and large the results are astonishingly close to what one would expect from a simple model in which IQ is largely inherited and environment has a small influence. We find, then, that all these different lines of evidence give results which are quite similar. They allow us to conclude that genetic factors account for something like 70 per cent (uncorrected figure) or 80 per cent (corrected figure) of individual differences in intelligence as measured by IQ tests.

There is one further type of evidence which is relevant here, and which has a profound practical importance—the so-called regression phenomenon. Regression to the mean has many important social consequences, and is rather a special phenomenon, so a discussion of it will be held over until Chapter 8. Let us merely note here that regression enables us to calculate heritability in a way that is independent of the methods so far described, and that the results are very similar to those arrived at by other methods.

THE BEST MODEL: A SIMPLE ONE

We may now summarise some of the major points discussed in this chapter. A very simple genetic and environmental model has been applied to numerous data from many different sources. This model postulates three sets of influences on IQ: genetic factors, home (between-family) factors and individual (within-family) factors. It does not postulate *statistical interaction* between IQ and environment—because none is needed.

This fact has often been the target of criticism, because, intuitively, interactions seem rather plausible. However, their presence to any degree would have made it impossible for the simple model to provide such a good account of the available data. The analysis of statistical interactions is a sophisticated statistical procedure too complex to enter into here. We should simply note that statistical interaction between heredity and environment would have marked effects on kinship correlations. Two individuals with both genetic make-up and environment in common would be subject to the same interaction between genes and environment and would therefore show an increased similarity. The effect would be most marked in MZ twins, who share all their genes. On the other hand, individuals who were adopted either share no genes (with their foster parents or foster-siblings) or share no environmental influences (with their parents or siblings or their separated twin): they will therefore interact uniquely and appear less alike.

As a result, an interaction between genetic factors and between-family factors would result in all the correlations for natural families being higher than the simple model would suggest, and all those for foster families being lower. The observed data show no such tendency. Interactions between genetic factors and within-family environmental factors may exist but would be difficult to detect. In view of the figures quoted, any such effects must be small indeed.

There is little evidence for interaction effects beyond a simple additive relationship. Although they may exist, the effects must be relatively small. We conclude, therefore, that a simple model giving a heritability of something like 80 per cent for IQ is both realistic and defensible. Errors of measurement are, of course, always present in scientific studies and make absolute accuracy impossible. But it seems very unlikely that the heritability of intelligence in modern Western countries would be lower than 70 per cent or higher than 85 per cent.

7
THE
INFLUENCE OF
ENVIRONMENT

Those who proclaim the importance of environment, or even completely deny the relevance of genetic factors, clearly have a duty to specify just what are the important environmental factors which they believe produce differences in IQ. And they have a duty to demonstrate that these factors actually produce such differences. It is, of course, much easier to manipulate the environment than to manipulate heredity, and one would therefore have expected a multitude of such studies demonstrating beyond doubt the influence of these factors. Actually, there is a dearth of such studies, and those that have been carried out tend to emphasise the relative lack of importance of environmental factors.

THE ENVIRONMENTAL LEVELLERS

Lawrence's orphanage study

To begin with, let us consider the interesting orphanage study of Lawrence. This study of children abandoned by their parents to an orphanage is important because any variation which might exist among such children should be due almost entirely to a biological factor, the genetic contribution of their true parents, because an orphanage provides as identical an environment for children as it is humanly possible to produce. If the contribution of genetic factors is really as important as is suggested by the studies reviewed in the last chapter, the variation in the IQ of orphanage children should only be slightly lower than that of a random sample of ordinary children brought up by their parents. If hereditary factors are relatively unimportant, or even non-existent, as Kamin has maintained, then there should be little if any variability among orphanage children. Lawrence found very little shrinkage in variation, and what he did find was virtually what would be expected with a heritability of 0.80 or 80 per cent. Unfortunately, the number of children in the study was not large enough to make the conclusions compelling, but as far as they go they strongly support a genetic, rather than an environmentalist, interpretation.

Studies of this kind are of particular importance from the social point of view because they indicate the limitations of egalitarian social policies in producing greater equality of IQ. It is impossible to think of any government, however powerful, that could provide a more equal environment for all its citizens than is produced in an orphanage, where the children all have the same living quarters, the same teachers, the same general environment, the same food, the same playmates and are, indeed, as far as humanly possible, treated in the most egalitarian manner conceivable. If under these conditions we still find almost as much variation in IQ as we do in the outside world, then clearly no government action can have much effect in this respect.

Social engineering in Warsaw

The extrapolation from a small orphanage study to large-scale social engineering may seem extravagant to some readers; fortunately we have direct evidence from a large-scale study by Anna Firkowska and her colleagues of the contribution of parental occupation and education to mental performance in 11-year-olds in Warsaw. The main purpose of the investigation was to separate out factors intrinsic to family social structure and position and factors extrinsic to it. Intrinsic factors include parental occupation and education, birth order and family size. Extrinsic factors include schooling, housing, health and welfare services, recreation, and criminality and employment rates.

In Warsaw there has been what the authors describe as "redress of inequalities of habitat among its people". Warsaw was razed at the end of the Second World War and rebuilt under a socialist government whose policy was to allocate dwellings, schools, and health facilities without regard to social class. Of the 14,238 children born in 1963 and living in Warsaw, 96 per cent were given the Raven's Progressive Matrices test and an arithmetic and a vocabulary test between March and June, 1974. The authors collected information on the children's families, and on the characteristics of schools in city districts. Parental education and occupation were used to arrive at a "family score".

Not unlike a capitalist society

Analysis showed that the initial assumption of even distribution was reasonable: members of the different social classes were distributed at random among the city's districts, and had identical educational and other facilities. It was found that mental performance was unrelated to school or district factors. But it did show a strong, even relationship with parental occupation and education—very much as it would in a typical capitalist society. The authors concluded that "an egalitarian social policy executed over a generation failed to override the association of social and family factors with cognitive development that is characteristic of more traditional industrial societies".

It is interesting to specify a little further the degree of equality achieved in Warsaw. Apparently, people of all levels of education and all types of occupation live in apartments that closely resemble each other, shop in identical stores that contain the same goods, and share similar catering and cultural centres. Schools and health facilities are equipped in the same way and uniformly accessible. Families of different occupation and culture live side by side in the same district, occupy buildings and homes of similar standard, and use the same schools and medical facilities. Yet this large-scale social engineering produces results very similar to those observed in the small-scale orphanage study. And both in turn strongly support the relative roles assigned to genetic and environmental factors in Chapter 6 on the basis of twin and family studies.

ENVIRONMENTS MATCHED

The Burks study

It is possible to study directly the specific effects of different environmental factors, such as the parents' income, the father's or mother's education and vocabulary, the level of culture of the home, or the number of books in the home library. Several such studies have been carried out following an investigation by Barbara Burks in 1928.

Table 2. Correlations between children's IQ and characteristics of the parents and home background. (Taken from Burks, 1928)

MEASURES		CORRELATIONS	
		FOSTER	NATURAL
FATHER'S	EDUCATION	0.01	0·27
	IQ	0.07	0·45
	VOCABULARY	0·13	0·47
MOTHER'S	EDUCATION	0·17	0·27
	IQ	0·19	0.46
	VOCABULARY	0·23	0.43
MIDPARENT IQ		0.20	0.52
CULTURE INDEX		0.25	0.44
WHITTIER INDEX		0.21	0.42
INCOME		0.23	0.24
HOUSE OWNERSHIP		0.25	0.32
NO. BOOKS IN HOME LIBRARY		0.16	0.34
PARENTAL SUPERVISION RATING		0.12	0.40
ESTIMATED MULTIPLE CORRELATION		0.35	0.53
ESTIMATED MULTIPLE CORRELATION CORRECTED FOR ATTENUATION		0.42	0.61

She took a great deal of trouble to match almost 200 foster families with 100 natural families on a number of potentially important factors such as parental intelligence and occupational status. The children were aged between five and 14. The home environment of all the families was assessed in some detail, including parents' interest in their children's welfare and education. A cultural index was arrived at by combining assessments of a number of factors—parents' education, how articulate they were, their spare-time interests, the quality of available reading material and evidence of artistic taste. An index of the material adequacy of the home, the Whittier index, was also used; this combines information on income, quality of food and home comforts, neatness, size of home and adequacy of parental supervision. Correlations between these and other measures of the home environment and the adopted children's IQ are shown in Table 2.

This pattern of correlations is interesting. Children's correlations with foster parents indicate the direct effect of the home environment. Correlations with natural parents are generally much higher and indicate the importance of underlying genetic factors unrelated in any direct way to the effects of the environment.

The greatest difference is in the extent to which a child's IQ correlates with his natural parents' or foster parents' IQ. It is, of course, much higher with natural parents (0.45 as against 0.07 in the case of fathers, 0.46 as against 0.19 in the case of mothers), which demonstrates the importance of heredity. The difference is least marked with economic factors such as income and home ownership, which appear to exert their influence almost entirely through the environment. The cultural quality of the home falls midway in its influence on IQ.

The foster home influence quantified

It is possible to calculate how much of the variability in the children's IQ was contributed by all the factors in the foster home environment, including foster parents' IQ. This is done by calculating the square of the correlation coefficient. Table 2 gives the correlation coefficient for the foster home factors as 0.42, the square of which is 0.18, or 18 per cent. This agrees well with the estimates of the effect of between-family environment given in Chapter 6. Other authors have found similar figures, although the most recent study of black children of similar age produced a much lower estimate.

The best and worst environments

Another way of looking at the issue of environment is in terms of the top and bottom 20 per cent of environments, that is, the best fifth and the worst fifth. Even such gross differences in social environment would be unlikely to produce differences in IQ larger than 18 points, compared with the 35 IQ points which differences in genetic endowment could be expected to produce. Nevertheless, 18 IQ points are far from negligible, and any changes in educational and social policies which could raise the bottom 20 per cent to the level of the top 20 per cent would be extremely valuable.

THE ENVIRONMENT ENRICHED

Heber's study

Is there any direct evidence of the contribution that environment can make to intelligence, for instance by increasing IQ? There are several studies, the most widely known perhaps being the one by R Heber which is still in progress but for which there are provisional results. Heber studied 40 children, selected at birth from a group living in the poorer areas of Milwaukee, a city with large numbers of seriously retarded children, mostly black. Half the children in the study were used as subjects in the experiment and half as controls. The control children took all the tests, but received no special treatment. The experimental subjects took part in an all-out effort, which lasted for several years, to improve their sensory, motor, language and thinking skills. From the age of three months onwards, for seven hours a day, five days a week, these children attended a university training centre for the mentally retarded, enjoying a planned, stimulating environment, and adequate medical care and nutrition. The mothers, too, were given an educational programme including home-making, child-rearing, and vocational training. The children were assessed every three weeks, either by standard tests or by tests of language and social development.

A 20-point gain

At the age of eight to nine, the experimental subjects had an approximate IQ of 104, while the controls averaged around 80. This is clearly an important gain, but several points should be noted. In the first place, the gain is no greater than would be expected on the basis of our genetic model; the children selected as controls were subjected to environmental influences well below the bottom 20 per cent (worst environment) used in our calculation, and the experimental children were exposed to an environment well above the top 20 per cent (best environment) in our example. Yet the observed figure of 104 is not higher than the maximum that would be predicted on the basis of our genetic model, and consequently Heber's results do not contradict it.

In the second place, there are many criticisms to be made of Heber's study. There are doubts about his method of matching the experimental subjects and the controls; too little detailed information is available on the study altogether; scores may well have been affected by the fact that the children were trained to answer the specific questions on which they were tested; and in any case IQs cannot be reliably measured at low ages. Furthermore, and most important, the children have not yet reached maturity, and until their final IQs at the age of 16 or so are known, we

Two home environments, two groups of children. Comfort and affluence tend to promote crystallised ability, which Western society values rather more than fluid ability. Would greater social equality, equal opportunity in education and housing reduce the environmental determinants of IQ?

cannot really say very much about the success of this experiment.

Other experiments have indicated that quite substantial IQ increases can be achieved with children suffering excessive deprivation and coming from unusually poor backgrounds, although these increases have never been greater than the genetic model would allow. A contribution by environment of about 20 IQ points on traditional IQ tests has occasionally been observed, but it does not run counter to our theory. It should be added that practically all the tests used in these studies have been tests of crystallised ability (drawing on acquired knowledge); it would be most interesting to find out how these methods of training and teaching affect performance on tests of fluid ability.

The factors involved in the Heber study, and other similar ones, are probably social and educational in nature, and it may be said in passing that deprivation of the severity he reported is rarely found in European countries.

THE ENVIRONMENT IMPOVERISHED

Malnutrition and IQ

Another source of deprivation often cited as a possible cause of low IQ is malnutrition, but it may be much less important than has been suggested. Consider, for instance, a study carried out in Holland by Stein and co-workers. They collected the test scores, at the age of 19, of some

20,000 Dutch army recruits whose mothers, during the German occupation, had been subjected to severe starvation in the crucial months around the time of the birth. These recruits showed no lasting general retardation when compared with 100,000 recruits whose mothers had not suffered starvation; the test used was the Progressive Matrices test. The study is important because the degree of malnutrition experienced by these mothers is exceedingly rare in Caucasian populations, and would never be found in any large groups. One may be justified in assuming that if such very severe food deprivation had no lasting effects on the IQs of the children involved, milder degrees of malnutrition would be equally harmless.

It has been suggested that the very severe and much more prolonged pattern of malnutrition often found in African children may have more pronounced effects, and some evidence for this has been provided by Stoch in South Africa. Be that as it may—and the evidence is by no means convincing, for it is difficult to demonstrate that mental retardation is a consequence of inadequate nutrition, rather than of the many other conditions that usually accompany it—nutritional factors cannot be said to play a role of any great importance in the IQ differences of European or North American children.

In conclusion, studies of specific environmental factors by and large provide quantitative support for the genetic model outlined in the last chapter. Environmental factors can be partly isolated and identified, and they have been found to affect IQ, but the size of their total effect is compatible with the hypothesis that 80 per cent of all the factors determining variance in IQ are genetic, and 20 per cent are environmental. Thus the studies reviewed in this chapter are complementary to those reviewed in the last, and point to much the same conclusion.

INTERPRETING THE CONCLUSIONS

It may be instructive at this stage to look at considerations involved in interpreting these conclusions, for they are frequently misunderstood and wrongly interpreted. In the first place, the results reported are relevant to *populations*, not to *individuals*. Heritability, in other words, is a population statistic. Because in a given population heredity accounts for 80 per cent of the variance in IQ, and environment for 20 per cent, it does not follow that these proportions would be the same for a given individual in that population, or in other cultures, or in the same culture at a different period in history.

As an example, consider England, or America. While the figures quoted give an approximate idea of the position as it is now, it is by no means certain that had these studies been carried out 200 or 300 years ago, results would have been the same. It seems quite likely that in those days environment played a much more important part than it does now, so that heritability would probably have been somewhat lower. Similarly, if the egalitarian policies of various Western governments continue to be pursued during the next century, bringing about greater equality in education, living conditions and so on, it seems quite possible that environmental determinants of individual differences in IQ would be reduced, and that the heritability of IQ would therefore increase.

Heritability is not God-given

It is important not to regard heritability as something God-given and universal. It applies to a given population, and is descriptive of that population. It is not prescriptive: current English or American figures will only ever be indicative of the particular population from which they are derived.

It should not be assumed that because differences in IQ are largely due to genetic factors, intelligence is fixed in some absolute sense, and that there is nothing that can be done about its level or its distribution. All that is said applies to conditions at a given time in a given place. Current environmental conditions in Western countries produce the results we have discussed. It is possible that new discoveries, either in physiology or in education, may alter conditions, and that in the new environment the population may achieve a different mean IQ, or a different distribution, or a different heritability. There is at present little sign of any such discoveries or inventions, and one may not be too optimistic that such inventions or discoveries are imminent. Nevertheless, the point must be made that in principle such possibilities cannot be ruled out. Everything in this book applies strictly to the here and now; the data do not enable us to make prophecies about the future.

Hebb's misleading analogy

Donald Hebb doubted the possibility of estimating heritability at all, and compared the effort to sort out the relative contributions of heredity and environment to obviously absurd efforts to sort out which is more important in deciding the size of a field—its length or its breadth. His analogy has been repeated innumerable times, but it is clearly inappropriate. By using the example of a single field, Hebb is implying that the geneticist attempts to sort out the influence of heredity and environment on a single individual; this would indeed be nonsensical. But the geneticist is concerned with a population. The question he asks is about the relative influence of genetic and environmental factors within that population. We must therefore rephrase Hebb: given a large number of rectangular fields, which is more influential in affecting differences in size between them, length or width, and is there any interaction between the two? That is a question which is quite easy to answer, using the statistical techniques known as analysis of variance; it may not be a very interesting or meaningful question, but it is certainly not nonsensical or unanswerable. The fact that Hebb and his many followers could completely misunderstand the whole basis of the genetic argument illustrates well the need for the inclusion of behavioural genetics among the subjects studied by psychologists.

School success and intelligence are not the same

A final word about education. It is sometimes said by critics that the intelligence quotient is in no way different from educational achievement. That this is untrue is clearly indicated by the fact that in studies of school achievement, genetic factors are shown to have far less effect on school achievement than on IQ scores.

Husen, for example, studied the records of the twins among all the males reporting for military service evaluation in Sweden at the age of 20. He used school records of achievement in arithmetic, writing and history for their final year of compulsory education, when the children were between 14 and 15 years old. The pattern of variation revealed a much lower heritability than for IQ, and considerable effects due to between-family environment. The same finding has been reached by many other authors in many different countries: achievement in school is very much due to intelligence, and hence has a genetic component, but it is influenced to a much greater extent than IQ by environmental factors, and the genetic component is therefore much smaller. There is ample evidence, then, that education and IQ are entirely different concepts, even though differences in educational achievement are largely determined by differences in IQ.

8

SPECIAL FACTORS: REGRESSION AND MATING SYSTEMS

The phenomenon known as the *regression effect*, or *regression to the mean*, can be seen in any organism which reproduces sexually, and in any trait which is less than 100 per cent inherited. It is simply the tendency for parents with extremes of a characteristic to produce less extreme offspring. Very tall parents will have children who are very tall but less tall than their parents. Very short parents will have children who are shorter than average, but taller than their parents. They will have regressed to the mean. If intelligence is heritable to the extent so far suggested, we would expect it to show regression to the mean.

The layman's misconception
What does this tell us? On the left of Figure 16 is the pattern most people will picture to themselves when told that differences in intelligence are 80 per cent inherited. There are 64 parents in all—four very dull, 16 dull, 24 average, 16 bright, and four very bright. Their children's IQs are distributed in much the same way, and the very dull parents have the very dull children, the very bright parents the very bright children, and so on along the line.

This picture, while intuitively appealing, is quite wrong, and is responsible for much of the opposition to the notion of the inheritance of intelligence. If things were really like this, then humanity would be inexorably divided into different classes—intellectuals and leaders at one end, hewers of wood and drawers of water at the other. No social mobility would be possible, and we would have, indeed, not a class system but a caste system.

However, because of regression, things are rather different, as is shown on the right of Figure 16. The offspring of the four very dull parents show very different IQs. Only one of the four children is very dull, two are dull, and one is average. Similarly, of the four children of the very bright parents, one is very bright, two are bright, and one is average. Of the children of average parents, one is very bright, one is very dull, six are bright and six dull, and only 10 are average. Thus of the four very bright

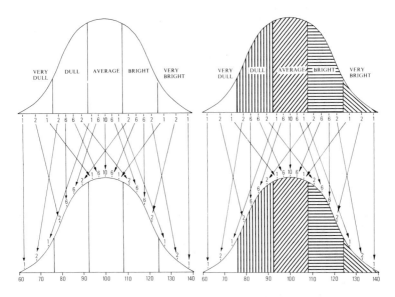

Fig. 16. Left: the layman's idea of the pattern of inheritance of IQ. It is mistaken because it disregards regression to the mean. Right: correct IQ inheritance

children, one has very bright parents, two have bright parents, and one has average parents. The diagram shows in detail just how these distributions are arranged. Regression redistributes genes in each generation, so that no caste system has a biological foundation (although of course it may be enforced, as in India, by social means).

Terman's gifted children

The regression phenomenon can be used to test the hypothesis of a specified degree of heritability for intelligence. Consider Lewis Terman's famous studies of gifted children. He selected 1,528 Californian children with an IQ of 140 or higher, and followed their progress to assess the importance of IQ in adult success and adjustment. The outcome was very positive: the great majority of the children were outstandingly successful academically, in business, and in the arts and sciences. This applied to females as well as males. In the cases where these gifted adults were unsuccessful, it could be shown that even at an early age they had shown psychiatric symptoms of neurosis, psychosis or other abnormalities. The mean IQ of Terman's subjects who married and had children was 152; that of their spouses was 125. (This is an obvious instance of the tendency, known as assortative mating, for bright men to marry bright women, to which we shall return shortly.) The mean IQ of all the parents was therefore 138.5, and that of 1,571 of their children was 133.2—evidence of some regression to the mean. Using the genetic formula for regression, we can calculate from the estimates of heritability (70 per cent) and within-family environmental variation (10 per cent) quoted in Chapter 6 that the children should be 33.5 points above the mean. In fact they were 33.2 points above the mean, very close to the prediction. Other studies on regression, starting at the lower end of the IQ scale, have produced comparable results, which reinforces our estimates of heritability.

Environmentalists are silent

Regression to the mean cannot be explained by those who believe that environmental factors are all-important. Very bright parents provide an optimum environment for their children; their homes usually offer books, study rooms and general cultural facilities. They provide educational drive and assistance, select the best available schools, obtain specialist teaching where needed, and generally try to push their children up the educational ladder. The atmosphere is favourable to learning, reading and general intellectual development. On the other hand, very dull parents provide the worst possible intellectual environment for their children, exactly the reverse of the bright parents' environment. If environmental factors are all-important, or even just very important, we would expect the children of very bright parents to do well—at least as well and possibly better than their parents—and the children of very dull parents to do poorly, perhaps even worse than their parents. But nothing of the kind happens. Instead, we find that the children of very bright parents show a *decrement* in IQ, and the children of very dull parents an *increment*. Environmentalists have been unable to say why.

Regression to the mean and social mobility

Regression is intimately connected with social mobility. In Western societies, only one person in three retains the social class of his or her parents. The major determinant of this upward or downward movement is IQ. When we look at the children of a given family, we find that the brighter ones rise in the social scale and the duller ones drop despite the same education, socio-economic status and home background in general. Thus regression mixes up the social classes, ensures social mobility, and favours meritocracy. At each mating the genes are shaken up and combined in a unique way, producing not only similarities but also differences between siblings and DZ twins. The results strongly determine the individual's life and career.

ASSORTATIVE MATING

We noted that in the Terman study there was some degree of assortative mating. Over and over again, other researchers have found the same trend, and we may accept it as fact that men and women who marry tend to be of similar intelligence.

Assortative mating may not seem very important. But it directly increases the genetic variability of a trait and the genetic correlation among relatives, at the same time as increasing differences between families.

An invaluable contribution to society

The importance of assortative mating for the genetic architecture of intelligence will be made clearer by some statistical calculations (they are based on the assumption that the current level of assortative mating in England and the United States has held for several generations). Assortative mating is shown to account for over half the IQs above 130, and four out of five of those over 145. With assortative mating there are approximately *20 times as many people with an IQ above 160* as there would be without assortative mating for intelligence. Jensen points out:

> "Such effects may greatly affect the character of a population in terms of intellectual resources. If our society were suddenly to engage in random mating with respect to intelligence, the intellectually most able of the next generation would not be as bright as the same upper percentage of the previous generation."

Of course, the percentage of mentally retarded would also be reduced, although perhaps to a somewhat lesser extent (about a quarter of mental retardation is attributable to rare genetic abnormalities and non-genetic causes such as brain injury and disease).

Assortative mating plays an enormous role, then, in producing those rare individuals with very high IQs whose contribution to society in science, the arts, politics, commerce and industry cannot be overestimated. Egalitarians who wish to promote a more equal society only have to marry dull wives (if bright) or bright wives (if dull), and to persuade others to do the same. By thus reversing assortative mating—and possibly producing negative assortative mating—they would drastically reduce the variability of IQ in the population.

INBREEDING DEPRESSION

Another important factor to take into consideration is the effect known as *inbreeding depression*. Inbreeding depression is the tendency for the offspring of marriages between blood relatives to be lower in various traits, including IQ, than the offspring of comparable parents who are not related. This occurs because high intelligence is genetically dominant over low intelligence, and in consanguineous marriages the recessive genes which lower intelligence have less of a chance of being offset by dominant ones. Recessive genes are more likely to pair up, thereby depressing intelligence.

Several large-scale studies of cousin marriages, mostly in Japan and Israel, have demonstrated this effect. In Israel, the rate for first-cousin marriages among Arabs was 4 per cent. For marriages between other cousins it was a very high 34 per cent—against 6 per cent in Japan and less than 1 per cent in Europe and America. Much stronger effects have been observed, of course, in the relatively rare cases where brother and sister, or father and daughter, produced viable offspring. These studies of inbreeding depression confirm that for many of the genes influencing IQ there is a marked degree of dominance. Our idea of the genetic architecture of intelligence is that much clearer.

How many genes are involved?

Finally, can we say anything about the number of genes that might be involved in the inheritance of intelligence? There are several ways of arriving at an estimate, all rather too technical for detailed discussion. One is based on the degree of dissimilarity in the IQs of siblings or of DZ twins. The larger the number of genes involved, the greater must the resemblance be.

Another method is to look at the relationship between the degree of inbreeding depression and the inbreeding coefficient, which tells us the degree of consanguinity between the parents. These and other methods lead us to postulate that roughly 50 genes are involved in the determination of differences in intelligence; this is a rough and ready estimate, but not one likely to be too far out.

INTELLIGENCE IN EVOLUTION

The general finding that high IQ is dominant over low IQ and that there is substantial variation in the dominant genes governing IQ makes sense when one realises that this kind of genetic control is characteristic of traits affecting biological fitness. Such traits have probably been subject to strong selection during the evolutionary process, and intelligence has no doubt played a major role. Exclusive stress on environmental factors does not take into account man's long evolution and the importance of intelligence in his development from ape-like ancestors. We are still far from having a completely accurate and satisfactory picture of the way that genetic factors determine our cognitive behaviour, but the rough outlines of the picture have emerged, and are not likely to be altered very much by subsequent research.

9
BIOLOGICAL MEASUREMENT OF IQ

So far, we have been concerned largely with statistical proofs—firstly with IQ tests and the items they are comprised of, which correlate with each other and provide us with "g", a useful measure of general intelligence, and secondly with studies of twins, adopted children, families, inbreeding depression and regression to the mean, all of which point to a strong genetic component in intelligence differences.

THE MOST CONVINCING PROOF

Surely, though, there must be underlying physiological reasons for these innate differences in ability. Recent work, some of it not yet published, has indeed begun to identify these physiological mechanisms and to measure them with considerable success. It provides the most convincing proof to date of the correctness of the genetic model of intelligence.

There are two approaches to this problem: the one taken by Arthur Jensen, and the one which I and my colleagues have taken. Jensen has been particularly concerned with the measurement of reaction times— the speed with which a person can react with a simple movement, such as pushing a button, to a simple stimulus, such as a light flashing on. He has been able to show that the hypothetical speed of nervous transmission measured in this way is quite highly correlated with intelligence as measured by traditional IQ tests.

My own interest has been rather in the brainwaves known as evoked potentials, as measured on the electroencephalograph (EEG). Evoked potentials tell us something about what is going on inside the brain when information is being transmitted. They also make it possible to frame theories about the nature of this transmission and its relationship to intelligence. As we shall see, it is closely correlated with intelligence.

JENSEN'S APPROACH: REACTION TIME TESTS

Experiments on reaction times usually have one of three formats, any of which can be used to establish the relationship between reaction time and intelligence.

Format 1: flashing lights

In the first of these the subject is presented with a console, as shown in Figure 17, which has a set of eight lights and eight buttons. When a light flashes on, he must at once turn it out by pressing the button associated with it. The interval between the light flashing and the button being pressed constitutes the reaction time.

Information theory: like a game of Twenty Questions

One, two, four or eight lamps may flash. The body of knowledge known as information theory—which deals with the way the brain processes information—tells us that every time the number of choices is doubled, we are adding one more "bit" of information. In technical terms, each bit of information equals the logarithm of the number of choices. In practical terms, the process is rather like the game of Twenty Questions, in which each question receives the answer Yes or No and you continue by a process of elimination.

You might say that the object in the reaction time experiment is to find out which light will flash. If there is only one lamp, no questions need be asked, and there is no bit of information involved. If there are two lamps, one question is sufficient—right or left? If there are four lamps, two questions are needed—for instance, odd or even-numbered, then right or left? This involves two bits of information. With eight lamps, we have three—the right set of four or the left set of four; odd or even-numbered; and then right or left? The important thing to note is that as the number of bits of information increases, so does reaction time, in regular fashion: each bit of information added increases the reaction time of a given subject by a set amount.

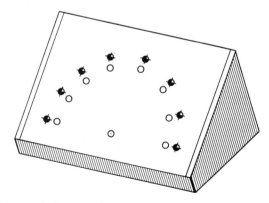

Fig. 17. Test console for measuring reaction time. (After Jensen, 1980)

Format 2: the probe type

Another format for reaction time experiments is to present the subject with a small set of digits (or letters), followed immediately by a single "probe" digit (or letter) to which he must respond by pressing a "Yes" or "No" button according to whether the probe was or was not included in the set. Reaction time increases as a linear function of the number of items in the set, and here again reaction times are quite highly correlated with intelligence.

Format 3: same or different?

In a third type of experiment, the subject is presented with stimuli which are the same or different, either physically (in appearance) or semantically (in meaning). As an example, the letters AA are physically the same, whereas Aa are physically different but semantically the same. Subjects are instructed to reply "same" or "different" to the stimulus. Here again, intelligent subjects respond much more quickly than dull ones.

Note that the mental processes involved in the different types of experiment are quite different. The flashing light experiment does not involve memory in any way. The probe type involves speed of scanning and short-term memory, while in the "same or different" format, because differences of meaning are involved, access is needed to long-term memory, where meaning is coded. Yet performance on all three correlates quite highly with IQ.

Variability, movement time and inspection time

Other useful measurements are variability of response, movement time and inspection time. If we test someone a number of times, we can measure the *variability* of his responses; he may be slow on some occasions and quick on others, while another person may be uniformly quick, or slow, or average. *Movement time* is the time which elapses from when the subject begins to move his hand from its resting place to when he presses the button. There is a great deal of evidence now to show that people with higher IQs on traditional intelligence tests have shorter reaction times, quicker movement times and less variable reaction times than people with lower IQs. The correlations for random samples of the population have reached around 0.5, a value which increases if more than one index is used.

Inspection time calls to mind George Santayana's saying that "Intelligence is quickness of seeing things as they are." In this type of experiment, the subject has to say which of two lines is the longer (the difference is quite appreciable). The length of exposure is so short to begin with that no accurate judgment can be made but is gradually increased until a judgment is possible. People differ in the length of time they require to make a correct judgment. This is called the *inspection time*, and it correlates quite highly with intelligence, longer intervals being required by people with lower IQs. It is difficult, here again, to see how education and cultural influences could affect the speed with which a person recognises such simple materials.

EYSENCK'S APPROACH: EVOKED POTENTIALS

With reaction time tests we are still in the field of psychology proper, although the assumption is that physiological mechanisms such as speed of transmission in neurons (nerve cells) are involved. A more direct, physiological, way of looking at the behaviour of the central nervous system is by studying the kind of brainwaves known as evoked potentials. The kind of electrical brain activity which the EEG traditionally charts is not very closely related to intelligence. Evoked potentials, on the other hand, are.

The Canadian psychologist J Ertl made use of the fact that a sudden stimulus, such as a flash of light or a sound delivered through earphones, gave rise to activity in the brain which registered as a characteristic set of waves on the EEG. Evoked potentials, as these waves came to be known, are measurable; but, unfortunately, interference makes it hard to arrive at a pure measurement (in technical terms, it is said that the signal-to-noise ratio is rather poor), so several waves have to be averaged to produce a measurable reaction.

A typical average evoked potential is shown in Figure 18. In this figure, A is the kind of EEG wave form found prior to the stimulus, B indicates the onset of the stimulus, while N and P are negative components (troughs) and positive components (peaks) of the averaged evoked potential. Most of the induced activity takes place within the

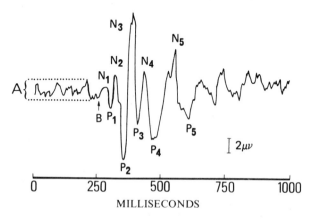

Fig. 18. Average evoked potential responses, showing EEG waves resulting from stimulus presented at point B. (Adapted from Shucard and Horn, 1972)

first quarter of a second or so and dies down over the next quarter or half second.

Slow, shallow waves—hallmark of dullness

Ertl discovered that dull subjects produced slower (more widely-spaced) waves than bright ones. This is quite clear in Figure 19—taken from Ertl's studies—which shows the average evoked potentials of a high, a medium and a low scorer on the Otis intelligence test. Figure 20 shows similar differences in the waves of 10 bright children (on the left) and 10 dull children (on the right), whose IQs were measured on the WISC, the children's version of the Wechsler test. The *latency* of the waves of dull children—in other words the interval between the waves— is clearly seen to be longer.

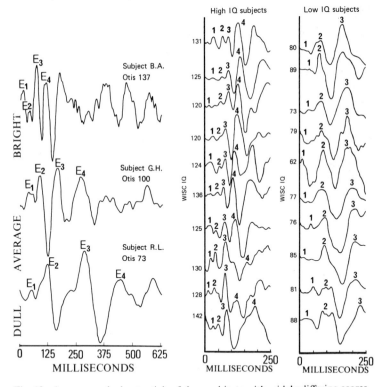

Fig. 19. Average evoked potentials of three subjects with widely differing scores on the Otis intelligence test. (Adapted from Ertl, 1968)

Fig. 20. Specimen visual evoked potentials for 10 high-IQ and 10 low-IQ subjects. (Adapted from Ertl and Schafer, 1969)

In our own laboratory Elaine Hendrickson found evidence to support this finding, as have other people before. However, her correlations were somewhat higher because she used auditory rather than visual stimuli (visual stimuli often produce distortions in EEG measurement). She also found differences in the *amplitude* of the waves of bright and dull subjects. Dull subjects produced shallower waves. By combining these two types of measurement—latency and amplitude—she was able to get correlations as high as 0·6 between measured intelligence and average evoked potentials.

In search of a theoretical basis

While all these findings were interesting and important, a proper theoretical basis for the use of evoked potentials was lacking. This was supplied by Alan Hendrickson, whose physiological and biochemical theory of intelligence and memory led to measurements resulting in substantially higher correlations with intelligence than either latency or amplitude, or both combined.

Hendrickson's theory of transmission errors

Hendrickson's suggestion was that as a message passes from one neuron to another through the cortex, the part of the brain involved in decision-making and higher mental functions, errors may occur. The greater the probability of an error occurring, the more difficulty a subject would have in solving cognitive problems. The evoked potential would reveal the number of errors occurring in transmission. An evoked potential, remember, is actually the *average* evoked potential of several transmissions. Errors in transmission, he postulated, would have the effect of smoothing out the wave, so that it would lose many of the squiggles and kinks characteristic of an error-free transmission. So long as there were no errors, the harder a problem, the more squiggles and kinks a wave would have. The waves characteristic of bright and dull people confirm Hendrickson's theory: dull people do have much blander waves than bright ones.

Squiggles and kinks translated

Hendrickson's theory also predicts that a measure of average evoked potential which looks at its complexity (its kinkiness, if you like) would correlate more highly with IQ than do traditional brainwaves, which, as we said earlier, relate poorly to IQ. Elaine Hendrickson tested the prediction, retrospectively by re-analysing previously published data, and prospectively by testing hundreds of adults and children for IQ and on the EEG. She found that correlations between evoked potential and IQ now shot up to higher than 0·8—in other words correlations between this psychophysiological measure and IQ were as high as those between one good IQ test and another. We now have direct evidence of important physiological factors closely related to cognitive functioning as measured by IQ tests. A concrete, measurable biological basis has been found for IQ.

New light on old controversies?

This finding opens up possibilities for solving all sorts of old problems and controversies. For instance, it should now be possible to measure directly the growth of intelligence in babies and young children, and the decline of intelligence with age. Differences between classes and races should be open to measurement along lines avoiding any cultural or educational contamination. So should the measurement of differences between the sexes.

It is no longer plausible to postulate a theory of intelligence which denies its biological foundation, or which assumes that observed differences in intelligence are due entirely or mainly to cultural, social and educational influences.

IO
RACIAL
AND CULTURAL
FACTORS

It is commonly believed that certain national, racial and cultural groups are more intelligent than others. Jews, Chinese and Japanese are often thought of as being particularly clever, Negroes and Mexican-Americans as being less able than average. There are two issues here which are often confused. The first question is whether in fact there are any differences in IQ between the various racial and national groups. This is relatively easy to establish. The second and much more difficult question is whether these differences are artifacts of testing, the result of cultural factors and the outcome of deprivation, or hereditarily determined and produced by genetic factors.

There is little debate about the actual existence of such differences: they have been demonstrated on quite large samples many times and seem to be very much in line with popular belief. The second question has not been answered with anything like the same degree of unanimity. This brief chapter will not go into great detail; it will simply state the facts of the case and leave interpretation to the reader.

RACIAL DIFFERENCES
Blacks: a 15-point lag
American blacks and American whites are the two groups most frequently studied—more is known about them than about all other racial groups combined. Figure 21 shows the distribution of IQ scores of a sample of black and white children tested in 1960; the mean scores are 80.7 (blacks) and 101.8 (whites). The black children in the sample came from the southern states of the US. Black children from the north usually score significantly higher, reducing the overall difference in IQ to something like 15 points. Black females usually score 3–4 points higher than black males (by contrast, among whites, as we have seen, there are no sex differences).

Scores on the Scholastic Aptitude Test (SAT), a test widely used in the US for selecting college students, show a similar pattern. The test has

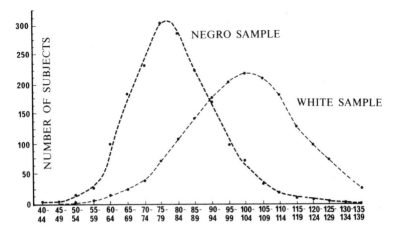

Fig. 21. Distribution of IQ scores of a sample of negro and white children. The mean scores are 80.7 (negro) and 101.8 (white). The negro children tested came from the southern states; children from northern states would have shown less marked differences. (After WA Kennedy et al)

two sections, one verbal, the other mathematical, with scores ranging from 200 to 800. In 1976–77, mean scores for high school students were 329 and 449 for blacks and whites respectively on the verbal part, and 355 and 490 on the mathematical part. Over the preceding five years, differences averaged much the same, with differences on the verbal part always slightly lower than differences on the mathematical part. This is a very carefully constructed test, which has been shown to have predictive accuracy and internal consistency very similar for the two races; in other words, it predicts scholastic success equally well for blacks and whites. The observed differences are pretty well what we would expect.

Figure 22 shows, in diagrammatic form, the distribution of IQs of whites, black males and black females in the total population. The mean difference is 15 points. Several features should be noted. In the first place, there is considerable overlap between the groups, so it is manifestly absurd to classify a person as bright or dull on the basis of his or her colour. The racist position of general white superiority is quite untenable: some blacks are greatly superior in IQ to many whites.

Whites are more extreme

At the extremes (very high and very low IQs) there is considerable disproportion. The line marked with an X in the figure shows the cut-off

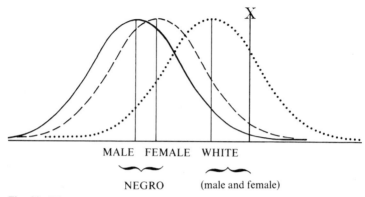

MALE   FEMALE   WHITE

NEGRO   (male and female)

Fig. 22. Diagrammatic distribution of IQs of whites, black males and black females. X marks the minimum IQ likely to be needed for college admission

point below which IQs would probably be too low for acceptance by a college or similar institution. Whites strongly predominate in this part of the diagram. Expressed in quantitative form, this means that at an IQ level of 70 or below, 16 per cent will be blacks and 2 per cent whites. At an IQ level of 100 or above, 16 per cent will be blacks and 50 per cent whites. At 115 or above (roughly the point where selective processes in secondary education are used to mark out "grammar school material" from the rest), 2 per cent will be blacks and 16 per cent whites. And at 130 and above (roughly the level of very good university students), 0.1 per cent will be blacks and 2 per cent whites, a ratio of 1 in 20.

Greater crystallised ability

But blacks, it is important to note, reverse the usual pattern in the US; they are the only racial group who do comparatively better on tests of crystallised ability than on tests of fluid ability, which suggests that their education has not handicapped them in relation to whites.

Studies using IQ tests have been carried out in Uganda, Jamaica, Tanzania, South Africa, Ghana and elsewhere, with similar results; blacks on the whole tend to have IQs between 70 and 80, even though many investigators selected children of higher than average socio-economic status and education, rather than random samples. Studies of black children in England have tended to give results comparable to those carried out in the US. There seems little doubt about the facts. Their interpretation, of course, is a different matter.

Japanese and Chinese outstrip whites

Mongoloid peoples—mostly Japanese and Chinese—have been mainly studied in countries, like the US, to which they emigrated, though some studies have also been carried out in places such as Japan and Hong Kong. The difficulty with studying the offspring of parents who emigrated is of course that they may not be representative of the native population: it is possible that those who emigrated were the most able and courageous, or the least conformist, or those who found it impossible to make a living because of low ability. Nevertheless, a clear picture of superiority emerges. Japanese and Chinese usually surpass whites on tests of fluid intelligence, but lag behind on tests of crystallised ability unless brought up in western-type schools. Chinese and Japanese born and brought up in the US outstrip whites on all tests of mental ability.

Jews do best of all

Jews have usually done better on IQ tests than any other group tested, both in the US and Great Britain. In one of the largest and best controlled studies of this type, carried out in Glasgow on a very representative sample of Jewish and gentile children, they emerged with a mean IQ of 118, boys and girls having very similar scores. If anything, this figure underestimates the mean IQ of the Jewish children, because those tested attended state schools, and of those not tested an unusually high proportion attended fee-paying schools where the mean IQ tends to be significantly higher. It is unlikely that these differences are attributable mainly to differences in socio-economic status: a London study found differences averaging 11 IQ points when Jewish and non-Jewish children of the same occupational background were compared.

Nobel Prizes and racial aptitudes

These differences in IQ agree well with the fact that an undue proportion of Nobel Prize-winners are Jews and that the publication *American Men and Women of Science*, which lists outstanding scientists, shows that in this field Jews outnumber non-Jews by something like 300 per cent. The Chinese did equally well in the physical and biological sciences, but less well in medicine or the social and behavioural sciences—although still surpassing non-Jewish whites. It is interesting to note that the Jewish contribution is particularly high in the more abstract sciences; Chinese are outstanding in the more observational and less abstract fields of earth sciences, botany, zoology, plant physiology and phytopathology, where they exceed Jews, who in these fields even fall behind non-Jewish whites. The educational and scientific achievements of Jews and Chinese in the US agree well with their superior performance on typical intelligence tests. Again, the facts are clear but their interpretation is debatable. Innate differences may play a part; so may greater stress on education in the family, and ambition born of suppression and racial intolerance on the part of the host race.

The British experience

So far we have talked about American or British whites and blacks, as if it could be assumed that these groupings were truly homogeneous. This is by no means so; as we have already pointed out, American blacks from the South usually produce lower IQs than do those from the North. Since most of the tests used were of the crystallised ability kind, differences in IQ may be the result of differences in education—after all, whites from the North also tend to do better than whites from the South. But even in a more homogeneous area like the British Isles, systematic differences among whites can be observed. Re-analysing large quantities of figures, Richard Lynn arrived at the distribution pictured in Figure 23. London and South-East England have the highest mean IQ score (102), and Ireland the lowest (96). This difference of 6 points is highly significant, from a practical as well as a statistical point of view. Lynn largely

Fig. 23. Mean IQs of standard regions of England, Wales, Scotland and Ireland. (After R. Lynn)

attributes these differences to selective emigration: the brightest Irish and Scots have tended to emigrate to England, and London in particular. He produces convincing evidence that over the past century this emigration pattern has changed the gene pool of Scotland from a position of potential superiority to one of actual inferiority. (Perhaps the newly discovered oil fields off the Scottish coast will do something to reverse the position.)

LOOKING FOR CAUSES

These, then, are some of the major facts about racial and cultural differences in IQ. Are these differences explainable in terms of educational differences, socio-economic status, poor nutrition, discrimination, racial prejudice, biased tests, white examiners testing coloured children, and other environmental variables, or must we postulate some innate differences? Different writers have come to different conclusions, Jensen, for example, arguing for hereditary causes, Kamin rejecting this possibility, and Vernon and others suggesting a "not proven" verdict.

The genetic approach
There are two entirely different ways of seeking a solution. The first is genetic studies of the kind outlined in Chapter 8 on twins, regression and inbreeding depression. Unfortunately, such studies are difficult, maybe even impossible, to carry out.

As I have said in my book *Race, Intelligence and Education*:

"The discovery of within-race genetic factors determining IQ differences is a *necessary*, but not a *sufficient* condition for accepting the genetic argument as applied to between-race differences. Can we go beyond this and argue that genetic studies . . . give *direct* support to the hereditarian position? The answer must, I think, be in the negative. The two populations involved (black and white) are separate populations, and none of the studies carried out on whites alone, such as twin studies, are feasible."

The genetic evidence is presumptive, not conclusive; on this point all experts are agreed.

The circumstantial approach
This leaves us with purely circumstantial evidence in support of environmentalist hypotheses, and the difficulties of evaluating circumstantial evidence are well known. This second approach, the circumstantial mode of proof, has already been used in the analysis of environmental factors in Chapter 7. We made deductions from the genetic and environmentalist hypotheses respectively, then looked at work which would provide evidence for or against our chosen hypothesis. Lawrence's orphanage study, or its social complement, the study of children brought up in egalitarian Warsaw, will serve as an example. On the environmen-

talist hypothesis these children should be very similar in IQ, as nearly all the hypothetical environmental factors supposed to produce IQ differences have been removed. But, in fact, we find that the differences are almost as large as before, which disproves the hypothesis.

We could carry out a similar procedure with regard to racial differences. Starting out with certain environmentalist hypotheses, we could search for evidence to support or disprove them. This is what Jensen has done in his book *Bias in Mental Testing*. Here we can only look at some representative studies to illustrate certain points.

MYTHS OF BIAS EXPLODED

Consider the argument that perhaps black children do worse than white children in IQ tests because testers are themselves white. This is easy to test; there are some 30 studies, some of which go one way, some the other. Overall, there is no evidence that the race of the tester makes any difference to the results of the test. The criticism can be ruled out of court.

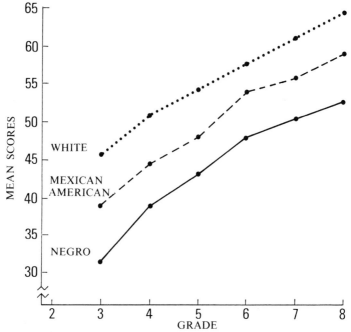

Fig. 24. Scores of white, Mexican-American and negro children on the Progressive Matrices test. (These are T-scores with a mean of 50.)

Fig. 25. Graph showing the relative standing of white, black and Mexican-American children from grades 4, 5 and 6 on four variables: socio-economic status, verbal IQ and school achievement, non-verbal (culture-fair) IQ, and rote memory. Scores are calculated in such a way that the children's standing on any one factor is independent of that on other factors. Note that when children are thus equated for ability and school achievement, Mexican-Americans are much the lowest in socio-economic status, with whites and blacks nearly equal. Conversely, with socio-economic status held constant, whites and Mexican-Americans are equal on culture-fair IQ, with negroes well below. With non-verbal IQ and socio-economic status held constant, whites are superior to both Mexican-Americans and negroes. These results suggest that Mexican-Americans are culturally deprived, and hence scholastically backward, but without any culture-fair IQ deficit. Negroes, on the other hand, show much less evidence of cultural deprivation, but much lower culture-fair IQs. (After A. Jensen)

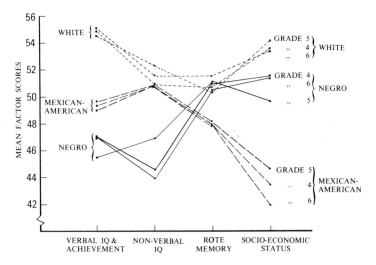

Language difficulties

Similarly, it has been argued that blacks are disadvantaged on "white" tests because of language difficulties; this can be tested by comparing results on verbal tests of crystallised ability and non-verbal of fluid ability. As we have said, this hypothesis does not stand up, because blacks actually do better on verbal than on non-verbal tests. Figure 24 shows the results of administering a purely non-verbal test (Progressive Matrices) to white, Mexican-American and negro groups; the differences are obvious.

Socio-economic status

Figure 25 shows the scores of these same three racial groups on verbal, non-verbal, and rote memory tests, as well as their socio-economic status; the results demonstrate the failure of the environmentalist hypothesis in graphic form. Figure 25 is also relevant to another argument, namely the alleged failure of black children to do well on IQ tests because of their low socio-economic status. As the figure shows, their socio-economic status is higher than that of the Mexican children, yet their IQ scores, both on verbal and non-verbal tests, are lower. Much the same argument can be applied to American Chinese; their socio-economic status is lower than that of whites, yet they do better than whites on non-verbal tests. This suggests strongly that socio-economic status may be much less of a determinant of IQ than many people have argued. The same finding has been made in Hong Kong: Chinese do better than whites, in spite of lower socio-economic status.

More directly relevant to the black-versus-white question are studies comparing black and white children whose parents had the same socio-economic status, who went to the same schools and lived in the same surroundings—in other words, all the allegedly relevant environmental factors were equal. A highly significant difference of some 12 IQ points was found. Matching subjects for socio-economic status makes some difference, then, but most of the IQ discrepancy persists.

Reaction times

Turning to performance on complex reaction time tests—which, as we have seen, correlate well with IQ—there is no question of socio-economic status, verbal knowledge or education being involved, yet there is a clear-cut difference in the speed of responses, whites being faster. Nor can it be argued that motivation might have been higher for the whites (another hypothesis frequently used to account for differences), for the two groups show similar speeds to begin with—before learning the required complex reaction pattern—and only diverge when they have learned the pattern. If there were differences in motivation, they would have been apparent from the beginning.

The gap is no narrower

Over the past 20 years blacks in the US have made strides economically and socially, even in the deep South. Discrimination and segregation have been enormously reduced by long-overdue government action and various Supreme Court rulings. Compared with conditions at the time of the First World War, the difference is even more dramatic. One would have expected these advances to reduce the IQ discrepancy between whites and blacks, but McGurk, summarising investigations into these differences, concludes that the gap in IQ has not been closed, nor even narrowed. "Intellectually," he writes, "the Negro of today bears the same relationship to the contemporary White as did the Negro of the

World War I era to the White of that time . . . . It seems clear that there has been no measurable improvement in either the absolute or the relative intelligence of the Negro."

SUMMING UP

A great deal more evidence of this kind could be adduced, but little useful purpose would be served. The environmentalist hypotheses which have been tested, such as those mentioned in the preceding pages, have been found wanting, which makes an environmentalist interpretation of racial differences less likely, and a genetic one more likely, to be correct. Of particular importance in this argument, as already pointed out, is the fact that non-white groups like the Chinese and the Japanese, even when tested in their own countries, do better on white IQ tests of the fluid intelligence type than do whites themselves. This is not compatible with the often-voiced objection that IQ tests are unfair to non-whites because they are devised by white psychologists. This argument is invalid in any case because, as was made clear in Chapter 3, test construction follows certain general rules which pretty well eliminate any subjective bias in favour of one's own racial group. That non-whites do better than whites on these tests is proof that they are fair in practice as well as in theory.

What can we conclude from this brief discussion of a very complex phenomenon? There is wide agreement that it would be premature to arrive at any very definite conclusions; at best we can argue about probabilities. Yet to deny that these probabilities point to a genetic basis for racial differences would be to disregard well-established facts. Environmentalists who wish to maintain their point of view are under an obligation to adduce better empirical material than is at present available to support their case. Simply protesting against the conclusion to which the facts seem to point is not enough. But whatever the final answer may be, we should remember Dr Johnson's reply when asked which sex was superior. Asked which man is more intelligent—a black man, a white man or a yellow man—we might reply, in Dr Johnson's vein: "Which black man, which white man, which yellow man?" Each person is an individual, not just a member of a race, group or sex, and should be treated as such.

11
SOCIAL CONSEQUENCES OF IQ MEASUREMENT

"It is likely that the mere fact of heritability in IQ is socially and politically important, and the more so the higher the heritability. Because the IQ measures something both *heritable* and *necessary for important social consequences, it cannot be dismissed* either *as an insignificant biological curiosity* or *as a wholly arbitrary cultural value. A mere biological curiosity it is not, because of its social predictiveness; a purely cultural artifact it is not, because of its heritability."*

RJ HERRNSTEIN

This is well said, and I would certainly consider it desirable if a public debate were to be initiated on just what these social consequences of IQ may be. What I have to say here constitutes merely one person's opinion and therefore differs in kind from the rest of this book. Where in previous chapters I have tried simply to transmit and interpret factual material, here I shall go beyond factual material to assess its relevance to social values and aspirations.

THE RECORD SET STRAIGHT

IQ tests do not *create* inequality

Let me begin by making clear a number of points. Some people seem to believe that social problems, such as racial inequality or social class differences, are created by psychologists and IQ tests. Nothing could be further from the truth. Social problems of this kind have always existed; IQ tests merely reduce them to a quantitative level, and thus make it possible to discuss them in a rational manner. Measurement never produces problems; it merely clarifies them. In Malaysia, the Chinese have IQs about 15 points higher than the Malays, who are the great majority. Differences in the abilities of the Chinese and the Malays have produced violent social reactions, including pogroms and murder. But IQ tests did not create the difficulties; they merely identified some of their origins.

Political beliefs and scientific standpoints

Another erroneous assumption is that political beliefs determine one's attitude towards the relative importance of heredity and environment. It is sometimes suggested that right-wingers favour genetic factors, left-wingers environmental factors. This is clearly untrue. I learned my genetics in large part from Professor JBS Haldane, who was not only one of the most gifted geneticists of the century, but also a leading member of the Communist Party of Great Britain and editor of its newspaper, *The Daily Worker*. As he made clear in his book *The Inequality of Man*, he was convinced of the importance of genetic factors as far as differences in intelligence are concerned, and did not believe that this fact was incompatible with communism. At the other end of the political spectrum, Professor JB Watson, founder of behaviourism, was an arch-conservative yet also the proponent of an extreme form of environmentalism. Many other examples could be given to illustrate the failure of agreement between political belief and a person's stand on the importance of genetic factors.

This mistake possibly arose because Stalin banned mental testing in 1935 on the grounds that it was "bourgeois"—at the same time as Hitler banned it as being "Jewish". But Stalin's anti-genetic stance, and his support for the environmentalist charlatan Lysenko, did not derive from any Marxist or Leninist argument. Indeed, both Marx and Lenin were firm believers in Darwin's doctrine of evolution and acknowledged the importance of genetic factors. One need only recall the communist manifesto: "From each according to his abilities, to each according to his needs." This clearly expresses the belief that different people will have different abilities, even in the communist heaven where all cultural, educational and other inequalities have been eradicated. Recent writings by communist psychologists behind the Iron Curtain make it clear that they are in agreement with this view, and indeed some of the most interesting recent work on the inheritance of cognitive abilities comes from Russia, East Germany and Poland.

Hitler: IQ tests would have shot his Aryan supremacy theories to ribbons.

Stalin: politically expedient to discount variations in ability in order to enforce equality.

An unwarranted leap of logic

A third error, equally as serious as the previous ones, is the assumption that certain facts outlined in previous chapters automatically lead to certain political conclusions. Thus it is sometimes said that because one racial group is superior to another in intelligence, the inferior group should be content to be relegated to simpler social tasks, while the superior group should assume all leadership positions. But this, of course, is nonsense: all racial and social groups that have been tested show great overlap in their abilities, so that, even if a criterion of intelligence were used in political matters, each person would still have to be judged on his own merits rather than simply as a member of his social or racial group. The fact that Japanese and Chinese score significantly higher on IQ tests than do Caucasian whites, whether in Europe or in America, does not lead us to suggest that leadership positions should automatically accrue to members of the yellow races, and that members of the white races should be reduced to inferior status.

Three youngsters: Jewish, Asiatic, Caucasian. IQ norms for their respective social and racial groups would not be predictive of their individual intellects or their potential as human beings.

Is intelligence overrated?

The fourth error, also quite common, is to exaggerate the importance of intelligence. Because one group is superior in intelligence to another, or because one person has a higher IQ than another, does not necessarily mean that the high-IQ person is more deserving of respect, or is more useful to society. There are many qualities, largely independent of intelligence, which are important, some perhaps more important than intelligence. Faith, hope and charity may be good examples. So might be "soul", or honesty, or hard work, persistence, kindliness, impartiality and a passion for justice, and many more besides. A person superior in intelligence may be a scoundrel, a psychopath or even a mass murderer; high intelligence is no insurance against low morality. The great villains of history, from Attila and Genghis Khan to Hitler and Stalin, have all been above average in intelligence; this does not make them admirable as people.

Intelligence, then, is only one of many qualities that make a person socially useful. However, there can be no doubt that it *is* an important quality, and that society as we know it depends very much on members showing a high degree of intelligence. Any complex and advanced society needs scientists, lawyers, doctors, engineers, politicians, artists and many others who show high intelligence; without them we could not exist. But equally, society could not exist or prosper without miners, bus drivers, labourers, dustmen, policemen or soldiers. A society composed entirely of Einsteins or Newtons would be as incapable of survival as would be one composed entirely of men and women with an IQ of 80 or 85. Division of labour is a hallmark of an advanced society, and this division largely takes place along intelligence lines.

POLITICAL CONSEQUENCES

Class divisions are inevitable

Certain consequences of a political nature do seem to follow from the facts outlined in this book. The marked differences in intelligence produced by genetic factors make it very difficult to see how any society could exist which did not subdivide into social classes, and indeed history shows that no society has ever existed which was not so divided. Even modern communist societies show class divisions which are at least as marked as those observed in capitalist societies, and often more so. But compensating for these class divisions is the social mobility produced by the genetic regression to the mean discussed in Chapter 8. Regression means that caste societies cannot have any enduring biological basis, and that descendants of middle-class or working-class families will share the chance of achieving a different social status from their parents.

To say this does not mean that it would be absolutely impossible for society, by social pressure, dictatorship or other means, to impose a caste system (as was the case in India), or to attempt to implement a classless system (any attempts at which have always failed). Science can only tell us what the facts are. It does not tell us what is desirable, although it may point out the difficulties of achieving what some people regard as desirable.

Discrimination: a two-edged sword

The facts outlined in this book may usefully throw some light on the vexed question of discrimination. It is sometimes said that discrimination exists whenever there is any departure from a precise quota of distribution between classes, races or the sexes in the number of places at universities, or in certain professions, or conversely in classes for the educationally subnormal (ESN). This would only be true if the different groups started with identical genetic equipment. However, as we have seen, people do not start out equally endowed. Children born of middle-class parents have higher IQs on the whole than children of working-class parents, in spite of regression; and if university places are awarded on the basis of students' intellectual promise, it follows that a higher proportion of middle-class than working-class children will go to university.

It is interesting to note that the proportion of working-class students going to university is higher in England than in Russia. In Russia, 53 per cent of university students come from clerical and professional backgrounds; in England the proportion is only 44 per cent. These figures should be looked at in terms of the proportion these classes constitute of the total population; in Russia, clerical and professional men and women constitute 5 per cent of the total population, whereas in England the professional and managerial classes constitute 14 per cent. Thus in Russia 5 per cent of the population is middle-class and contributes 53 per cent of university students, whereas in England 14 per cent of the population is middle-class but contributes only 44 per cent of students. In France, Germany and Scandinavia the figures are similar to the English ones; in America the working-class contribution is even higher.

Quotas of any kind are discriminatory

Much the same argument could be presented with respect to race. It has often been observed that Jewish, and in America Chinese and Japanese, immigrants obtain a disproportionate number of university places. This is not evidence of racist policy on the part of university administrations; it is a simple consequence of the higher IQs of Jewish, Japanese and Chinese adolescents. These facts of racial superiority have often led in the past to discrimination in the form of quotas limiting the admission of Jews or Japanese and Chinese. It need hardly be said that discrimination of any kind is alien to the democratic ethos and should not be tolerated in a society devoted to racial equality. The major tenet of non-discrimination is surely that each person should be judged as an individual, not as a member of a racial, religious or any other kind of group. Any attempt to establish quotas violates this principle.

How, in terms of this principle, are we to look upon the decision in 1979 by Robert Peckham, a federal district court judge in San Francisco, in the Larry P versus Riles case? Peckham declared that the use of standard IQ tests to place black children in classes for the retarded violated not only the California constitution but also the 14th Amendment of the US constitution which guarantees equality of protection. The court accordingly ordered California schools, and the psychologists who work in them, to observe a ban on IQ tests; school districts were also ordered to move quickly to reduce racial imbalance in classes for the retarded. (Similar moves have been suggested in England to reduce the proportion of West Indian children in educationally subnormal classes.)

The answer to Judge Peckham, who achieved immortality by joining Stalin and Hitler in banning IQ tests, is that the differences in achievement and ability quantified by such tests are not created by the tests, and any attempt to treat the problem on a quota basis is basically racist and disregards the rights of the individual to be treated, not as a member of a race but as an individual.

ESN classes (EMR classes in America—for the educable mentally retarded) were organised originally to provide a service for children who could not keep up with the academic work in ordinary classes; attempts were made to bring them up to standard, if possible, enabling them to have a more successful educational career than otherwise. The introduction of quota systems simply deprives very low IQ children of an opportunity for much-needed help, and creates great difficulties for the schoolteacher, who finds it impossible to teach them in ordinary classes without detracting from the education of the other children. The consequences of what is no doubt a well-meant judgment are likely to be disastrous for the children affected by it, and may seriously affect the whole school system. This is more than a forecast of things to come: such bans in the United States have produced these results with sad and monotonous regularity. Whatever the answer to racial and class problems may be, this is not it.

A call for reason

It is unfortunate that there has been very little public discussion of the social and political implications of major findings in the area of intelligence. The issues raised are profound and important, but so far we have only witnessed an unholy war of words, with extremists denouncing each other as "fascists" or "communists", "racists" or "nigger lovers". Emotions have run very high indeed, and those who have drawn attention to the genetic role in IQ and other differences have been accused of following in the footsteps of Hitler, and of seeking genocide. This, of course, is an absurd attempt to establish guilt by association. The same smear tactics could be used to "prove" that socialism is a vile and evil creed. Was not Hitler's party the National-*Socialist* Party, and did not his party programme call for the same kind of socialist measures as the Labour Party of Great Britain? Such "proofs" are highly dangerous.

On the one hand, the debate deals with academic facts. On the other, it touches on important social concerns which involve, as well as these facts, ethical and moral issues beyond the range of empirical research. The issues are important and may even be vital to the survival of a democratic society. They should be discussed calmly and rationally, not with emotional diatribes and name-calling. It is to be hoped that the debate carried on in these pages may help to define the issues and enable the reader to draw his own conclusions.

12
SOME
HISTORICAL FACTS
ABOUT IQ TESTING

"If ... the impression takes root that these tests really measure intelligence, that they constitute a sort of last judgment on the child's capacity, that they reveal 'scientifically' his predestined ability, then it would be a thousand times better if all the intelligence testers and all their questionnaires were sunk without warning in the Sargasso Sea." WALTER LIPPMANN, 1922

BINET AND THE EARLY TESTERS

The first widely used intelligence test was created in France, in 1905, by Alfred Binet. The public school authorities in Paris had asked Binet to devise a method that might pick out in advance those children who were not likely to learn much from the teaching methods and curriculum of ordinary schools. These children could then be placed in special classes.

The test pieced together by Binet put different sets of questions to children of different ages. The questions depended on the child's general fund of knowledge, and some were intended to measure how well the child could reason and how sound his judgment was. The basic idea was that, on average, older children are able to answer more difficult questions than younger children. Thus any given child could be assigned a "mental age", depending upon what questions he could answer. Pierre, for example, would be given a mental age of eight if he could answer questions passed by the average eight-year-old, but could not answer questions passed by the average nine-year-old. Whether Pierre was said to be retarded, average or bright depended upon the relation between his mental age and his chronological age. Thus an 11-year-old with a mental age of eight was clearly retarded, but a five-year-old who could answer the same questions was obviously bright.

To Binet's great satisfaction, performance on his brief test correlated with teachers' judgments about which children seemed bright in school and which seemed dull. The fact that test scores were related to success at school work was thought to demonstrate that the test in fact measured "intelligence". This relation, which depended upon Binet's use of school-like questions, is what made his test more useful and more influential than the so-called "mental tests" with which earlier psychologists had experimented.

Galton and the eugenics movement

Earlier interest in mental tests had stemmed largely from the work in the 1860s of Francis Galton, who founded the eugenics movement. Galton believed firmly in the inheritance of mental ability and of just about everything else. The purpose of eugenics was to improve the human breed by encouraging the genetically superior to have many children, and by discouraging (or preventing) the genetically inferior from reproducing at all. To accomplish such a result, however, it would be necessary to devise tests and measurements that could identify the genetically superior and inferior. Hence the interest of Galton and his followers in measuring physical and psychological differences between individuals and between races.

The earliest "mental testers", following Galton's lead, concentrated on obtaining precise measurements, preferring tests of the kind used in laboratories to the kind used in schools. Laboratory tests make it possible, for example, to determine a person's reaction time to a fraction of a second by measuring how long it takes him to press a telegraph key in response to the sound of a buzzer. To the early experimenters it seemed reasonable that quickness in such simple "mental reactions" might be related to "quick-wittedness" in general, or to "intelligence". It soon became apparent, however, that precisely measured performances in such laboratory tasks did not even correlate with each other—far less with school grades, or other assumed indices of intelligence. The experimental tests inspired by Galton's interest in eugenics came to a dead end. But Binet, whose motives were practical and humanitarian, provided the Galtonians with fresh ammunition.

BINET'S IDEAS MISUSED

The IQ test, in Binet's view, was not a measure of "innate" or "inborn" intelligence. Binet thought of his test as a diagnostic instrument which made it possible to pick out children whose intelligence was not developing properly, who could then be given courses in what he called "mental orthopedics". The point of such courses was to *increase* the intelligence of children who had scored low on IQ tests. Binet's attitude is clear: he firmly rebuked those who believed that "the intelligence of an individual is a fixed quantity, a quantity that one cannot augment.... We must protest and react against this brutal pessimism."

Early racism

Those who first translated and used Binet's test, both in the United

States and in England, were convinced Galtonians, however. They *knew*, even before data had been collected, that intelligence had to be largely hereditary. Thus Lewis Terman, who introduced the Stanford–Binet test to the United States in 1916, wrote that IQs in the 70 to 80 range were "very, very common among Spanish-Indian and Mexican families of the Southwest and also among negroes". He continued:

> "Their dullness seems to be racial, or at least inherent in the family stocks from which they come. ... The whole question of racial differences in mental traits will have to be taken up anew. ... The writer predicts that when this is done there will be discovered enormously significant racial differences in general intelligence, differences which cannot be wiped out by any scheme of mental culture.

> "Children of this group should be segregated in special classes. ... They cannot master abstractions, but they can often be made efficient workers. ... There is no possibility at present of convincing society that they should not be allowed to reproduce, although from a eugenic point of view they constitute a grave problem because of their unusually prolific breeding."

There was no doubt in Terman's mind that differences in the IQ scores of different racial groups were produced by genetic differences between the races. And IQ differences *within* a particular racial group were also determined by genes. Terman believed that members of the upper social and economic classes possessed superior genes, which they passed on to their children. The same point of view was clearly expressed by another early translator of Binet's test, Henry Goddard in 1920. "The fixed character of mental levels", Goddard argued, caused the unending plight of the degenerate poor and of the unemployed. This "fixed" mental level was said to be measured by Binet's test—a view entirely opposed to Binet's own.

Black and brown-skinned races have fared consistently badly at the hands of eugenicists and the politicians who put their theories into practice.

In England, the early mental testers made extravagant claims about the hereditary basis of test performance even before they became acquainted with Binet's test. As early as 1909 Cyril Burt administered a set of crude tests to two very small groups of schoolchildren in the city of Oxford. The children at one school were the sons of Oxford dons, Fellows of the Royal Society and such like, while at the other school they were the sons of ordinary townspeople. Burt maintained that the children of higher social class did better on the tests—and that this demonstrated that intelligence was inherited. By 1912 Burt could write that "the evidence is conclusive" for the inheritance of mental capacities. The fact that parents provide children with their environments, as well as with their genes, seems to have made no impression upon Burt, or upon Terman and Goddard.

Sterilisation laws

The uncritical belief in the power of heredity, linked to the advocacy of eugenic ideas, was already widespread when Binet's test appeared. More than 30 American states followed the lead taken by Indiana in 1907 in passing eugenic sterilisation laws which provided for the compulsory sterilisation of, among others, criminals, idiots, imbeciles, epileptics, rapists, lunatics, drunkards, drug fiends, syphilitics, moral and sexual perverts, and "diseased and degenerate persons". The laws declared as a matter of legal fact, that the various defects of all these offenders were transmitted through the genes. The wholly unscientific fantasies of the eugenicists encouraged the naive claim that sterilisation of offenders would eliminate these undesirable traits from the population. Fortunately, the sterilisation laws were not often enforced. When they were, the victims were poor.

Immigration quotas

In the hands of eugenicists like Henry Goddard, the new science of mental testing was also employed to reduce unwanted immigration into the United States by the peoples of southern and eastern Europe. Goddard administered Binet's test in translation, together with some "non-verbal" or "performance" tests, to a number of "average immigrants" arriving at New York. His results claimed to show that 83 per cent of Jews, 87 per cent of Russians, 80 per cent of Hungarians, and 79 per cent of Italians were "feeble-minded". There was no doubt in Goddard's mind—or in the minds of other American mental testers—that tests producing such results measured "innate ability".

This naive belief had far-reaching consequences. During the First World War, the American army administered the new mental tests—basically modifications of Binet's pioneer procedures—to literally millions of men. After the war the National Academy of Sciences published the average scores of immigrant soldiers from different European countries. The highest scorers were immigrants from England,

Scotland, Canada and Scandinavia, the lowest from Russia, Italy and Poland. Mental testers concluded that "Nordics" were genetically superior to the "Alpine" and "Mediterranean" races. The claim was confidently repeated, this time by Brigham and others, that the tests measured "native, inborn intelligence". The Army data were cited repeatedly in congressional and public debates which led to the passage in 1924 of the overtly racist "national origin quotas" designed to reduce immigration by the genetically inferior peoples of southern and eastern Europe.

The educational scrap-heap

The IQ test has also played an important part in the American school system—especially in assigning lower class and minority children to dead-end classes for the "educable mentally retarded". The fact that a child has a low IQ score has been misinterpreted to mean that the child does not have the capacity to learn school subjects. The IQ test played an even more central role in England, where it formed the basis for the selective education system introduced after the Second World War. On the strength of Cyril Burt's enthusiastic argument that a test given to a child at the age of 11 could measure its "innate intelligence", it was decided to use the results of tests administered to 11-year-olds to "stream" children into one of three separate—and far from equal—school systems.

"Intelligence", Burt wrote in 1947, "will enter into everything the child says, thinks, does or attempts, both while he is at school and later on. ... If intelligence is innate, the child's degree of intelligence is permanently limited. No amount of teaching will turn the child who is genuinely defective in general intelligence into a normal pupil." This pessimistic claim—so antithetical to Binet's point of view—was later put into even plainer language when Burt equated intelligence with "educable capacity". "Capacity", he stated in 1961, "must obviously limit content. It is impossible for a pint jug to hold more than a pint of milk; and it is equally impossible for a child's educational attainments to rise higher than his educable capacity permits." In other words, an IQ test could measure a child's capacity for education, and it was obviously nonsensical to try to force more education into the child's head than could be fitted in, as indicated by his score.

The notion that a so-called intelligence test can somehow measure innate "capacity" or "potential" was considered and explicitly rejected in 1975 by a committee of testing experts appointed by the American Psychological Association's Board of Scientific Affairs. The Cleary committee declared:

> "A distinction is drawn traditionally between intelligence and achievement tests. A naive statement of the difference is that the intelligence test measures capacity to learn and the achievement test measures what has been learned. But items in all psychological and educational tests measure acquired behavior. ... An attempt to

recognize the incongruity of a behavioral measure as a measure of capacity is illustrated by the statement that the intelligence tests contain items that everyone has an equal opportunity to learn. This statement can be dismissed as false. . . . There is no merit in maintaining a fiction."

Politics and the nature-nurture debate

The points made by the Cleary committee seem so obvious that it is hard to understand how any psychologist could believe that IQ tests measure innate intelligence. Perhaps we should look at a scientist's social and political beliefs, for they are likely to influence the way he interprets IQ data. Pastore has shown that eminent scientists who stressed the "nature" side of the nature–nurture controversy tended to be politically conservative, while those who stressed the "nurture" side tended to be liberal.

We have seen that the pioneers of IQ testing in the United States were enthusiastic advocates of eugenic policies, and believers in the innate basis of IQ test scores, even before they collected data. The 1903 notebook of Cyril Burt, then a 20-year-old Oxford undergraduate, contains the following neatly handwritten entry:

> "The problem of the very poor—chronic poverty: Little prospect of the solution of the problem without the forcible detention of the wreckage of society . . . preventing them from propagating their species."

With beliefs of that sort, it is not surprising that Burt could interpret the fact that slum children did poorly on Binet's test as a sign of their genetic inferiority—and as proof that the test miraculously measured inborn ability.

THE HEREDITARIAN ARGUMENT

There are, of course, a number of facts cited by hereditarians to support their claim that IQ is largely determined by the genes. To begin with, it is clear that IQ scores tend to run in families. Parents with high IQs tend to have children with high IQs, just as parents with low IQs tend to have low-IQ children. The closer the biological relationship between two members of a family, the more they are likely to resemble each other in IQ. Children of different socio-economic classes have different average IQs. Children of manual workers tend to have lower IQs than children of professors and executives—a fact that has convinced some professors that they are genetically superior to manual workers. To some theorists, the fact that blacks in the United States have a lower average IQ than do whites is still further evidence that tests must be measuring inborn ability.

The most recent wave of interest in the genetic basis of IQ was largely provoked by concern over racial questions in the United States. Professor

Arthur Jensen argued in an influential article in 1969 that American "compensatory education" programmes—aimed primarily at improving the scholastic performance of poor black children—had not worked. The failure of such programmes was, in his view, inevitable, for the data of Cyril Burt, described by Jensen as "the most satisfactory attempt" to measure the heritability of IQ, had indicated that about 80 per cent of the variation in whites' IQs was genetic. It was plausible to suppose, therefore, Jensen argued, that the difference in average IQ between blacks and whites was caused by the genetic inferiority of blacks. Finally, the argument went, differences with a highly heritable basis could not be eliminated by environmental treatments such as compensatory education.

Fallacious logic

The pages that follow will examine critically the evidence used to demonstrate the high heritability of IQ among whites. It is extraordinarily weak. Indeed, what was thought to be the clearest evidence—Burt's—is now recognised to be fraudulent. We should note at the outset, however, that even if the claim that IQ is highly heritable among whites were true, the remaining steps in Jensen's argument are entirely fallacious. Though it may seem intuitively correct to assert that a highly heritable trait cannot be changed by environmental treatment, it is simply not the case. Weak eyesight, for example, may be highly heritable, but it is easy to correct with spectacles, and we do not regard an eye test as measuring some fixed and unchangeable "capacity to see". And take the case of phenylketonuria, a rare form of extreme mental retardation which is caused by the inheritance of a single gene. The defective gene results in a metabolic defect which in turn affects development of the brain and nervous system. Yet it is simple to prevent mental retardation from occurring in a child born with the gene by feeding it a special diet with as little phenylalanine as possible. There is no reason, then, to believe that the role of genes—whatever it may be—in *producing* a trait is in any way related to the ease (or difficulty) of *modifying* that trait by environmental methods.

THE CONCEPT OF HERITABILITY

There is an unfortunate tendency for many readers—and for some scientific writers—to misunderstand the technical concept of "heritability". To assert that the heritability of IQ is 0.80 is *not* to assert that 80 per cent of John Smith's IQ is inherited, while 20 per cent is produced by environment. Rather, it is to claim that—in some particular population, at some point in time—about 80 per cent of the variation in IQ, or IQ *differences* among individuals, is determined by genetic differences. Note, for example, that the heritability of two-eyedness in human populations is close to zero. That does *not* mean that the possession of two eyes is not determined by our human genes. What it means is that there is very little *variation* among us in the number of eyes we possess, and that any such variation is not related to individual genetic differences. The vast majority of people with only one eye, or none, have lost eyes through environmental accident, and not through transmitted genetic defect.

The heritability of a trait in a human population is, to say the least, very difficult to estimate, some would say impossible. When an estimate is made, it applies at best to a particular population at a particular time. The heritability of the same trait may be very different in other human populations, or in the same population at later (or earlier) times. The heritability of a trait is not some "law of nature". It is a population statistic, rather like the death rate in Madagascar during the fourth century—which tells us nothing about the death rate in North America today.

The elementary confusion in Jensenism

Finally, it is important to realise that even if the heritability of a trait is high *within* each of two populations, that in no way allows us to conclude that a difference in the average value of the trait *between* the two populations is genetically caused. This elementary confusion lies at the root of what the *New York Times* christened "Jensenism". The basic claim by Jensen was that the "fact" of high IQ heritability *within* both the white and black populations made it likely that the 15-point difference in average IQ *between* the two groups was caused by the genetic inferiority of blacks. The fallacy in this claim—even if Jensen's alleged "fact" were true—has since been pointed out by many geneticists and psychologists. The fallacy can be made obvious by a simple example.

We fill a white sack and a black sack with a mixture of different genetic varieties of corn seed. We make certain that the proportions of each variety of seed are identical in each sack. We then plant the seed from the white sack in fertile Field A, while that from the black sack is planted in barren Field B. We will observe that within Field A, as within Field B, there is considerable variation in the height of individual corn plants. This variation will be due largely to genetic factors (seed differences). We will also observe, however, that the average height of plants in Field A is greater than that in Field B. That difference will be entirely due to environmental factors (the soil). The same is true of IQs: differences in the average IQ of various human populations could be entirely due to environmental differences, even if *within* each population all variation were due to genetic differences!

The following pages will demonstrate that many of the key "facts" asserted by Jensen, Eysenck and other hereditarian IQ theorists are simply not true. Perhaps more important, it should be clear at the outset that even if the asserted facts were true, the implications drawn from them do not follow logically. We are entitled to conclude that today, as in the past, untrue facts and fallacious conclusions tend to reflect the social and ideological biases of the theorists.

13

THE
CYRIL BURT
AFFAIR

"I could only wish that modern workers would follow his [Burt's]
example . . ." HANS J EYSENCK, 1974

For many years, the central evidence cited to support the claim that IQ
is a highly heritable trait was the massive life's work of the late Sir Cyril
Burt. The importance of Burt's work is difficult to exaggerate. The
famous Jensen article of 1969 leaned heavily on Burt's work, which it
described as "the most satisfactory attempt" to estimate the heritability
of IQ. When Burt died, Jensen described him, in 1972, as "a born
nobleman", whose "larger, more representative samples than any other
investigator in the field has ever assembled" would secure Burt's "place
in the history of science". Hans Eysenck indicated that he drew "rather
heavily" on Burt's work, and cited "the outstanding quality of the design
and the statistical treatment in his studies".

CLEAR-CUT RESULTS

The impact of Burt's data was so great because, if taken at face value,
his results seemed entirely clear-cut. The Burt studies provided apparently
satisfactory answers to almost every conceivable objection. For example,
a theoretically simple and powerful way of studying the heritability of IQ
is to measure the IQ correlation of pairs of identical twins who have been
reared apart from one another. Pairs of identical twins, of course, have
identical genes. When such twins have been reared apart, they
presumably have only their heredity—and not their environment—in
common. If such twins resemble one another in IQ, then it must be due
to the only factor they have in common, heredity. This logic holds,
however, only if we can be sure that the environments in which the
separated twins were reared did *not* resemble one another.

To find identical twins who have been reared apart is no easy matter.
There have been only four reported studies of such twins. The largest of
the studies—purportedly based on 53 pairs of separated twins—was

reported by Cyril Burt in 1966 and claimed to observe a higher IQ correlation than that reported by other investigators. The most important virtue of his study, however, was one that Burt and those who cited his work stressed repeatedly: this was said to be the only study to attempt any systematic or quantitative measurement of the environments in which separated twins were reared. The socio-economic status (SES) of the households in which Burt's separated twins were reared was rated on a six-point scale. Although there was *no* correlation *at all* between the SES of the homes in which separated twins had grown up (which could be considered extraordinary), the twins nevertheless resembled one another greatly in IQ. This appeared to be powerful evidence indeed for the heritability of IQ.

Quite apart from his twin studies, Burt also contributed enormous quantities of data correlating the IQs of biological relatives of varying degrees of closeness. There are some categories of relatives—for instance second cousins, uncle–nephew and grandparent–grandchild—for whom the *only* reported IQ correlations are those reported by Burt. The only investigator who ever claimed to administer the same IQ test, in the same population, to all the different categories of blood relatives was Cyril Burt. The results were again extraordinarily clear-cut: the closer the biological relatedness, the higher the IQ correlation. The Burt data on relatives and on twins were routinely cited in textbooks of psychology, genetics and education as clear evidence of the high heritability of IQ.

ELEMENTARY FLAWS

With hindsight, it seems almost incredible that Burt's data could ever have been taken seriously. To begin with, Burt never provided even the most elementary information about how, where or when his purported data had been collected. When a scientist reports results, it is essential that he provide a clear and reasonably detailed account of the procedures he employed in obtaining the results. This was never done by Burt. Incredibly, in most of his papers there is not even any information about which IQ test was supposedly used to obtain the reported correlation.

Vagueness about method
The first large collection of IQ correlations among relatives was reported by Burt in 1943. The paper contains virtually no information about methods or procedure. The alleged correlations are merely presented, without supporting details. The only reference to procedure is the following: "Some of the inquiries have been published in LCC [London County Council] reports or elsewhere; but the majority remain buried in typed memoranda or degree theses." When scientists refer to primary sources and to documentation, they do not usually cite "elsewhere" as the place where something has been published. They do not tend, when talking about genuine work, to emphasise that the work

is "buried" and inaccessible. The reader should not be surprised to learn that none of the London County Council reports, typed memoranda or degree theses vaguely referred to by Burt in the cited sentence has ever come to light.

The fact that Burt had worked as a school psychologist for many years made it reasonable to suppose that IQ test scores of children were easily available to him. But where and how did Burt obtain IQ test scores for adults? In a single paper in 1956, Burt and Howard reported intelligence correlations based, among others, on 963 parent–child pairs, 321 grandparent–grandchild pairs and 375 uncle–nephew pairs. Yet there was no reference in this paper to the procedures employed; according to Burt, they had already been described in his earlier papers.

"Camouflaged" interviews

There is, in fact, a telltale footnote in one of the earlier papers. With respect to a reported correlation between parent and child, Burt wrote in that footnote in 1955: "For the assessments of the parents we relied chiefly on personal interviews; but in doubtful or borderline cases an open or a camouflaged test was employed." That is, in assigning intelligence scores to adults, Burt did not even *claim* to have administered an objective, standardised IQ test. There was no description by Burt of which "open" IQ test might sometimes have been employed. The idea of Professor Burt administering an occasional "camouflaged" IQ test to grandparents and uncles while interviewing them might have merit as comic opera—but as science it is absurd. This work, however, was cited as "the most satisfactory attempt" to estimate the heritability of IQ. That surely tells us something about the scientific calibre of work in this area, or about the critical standards of authorities in this area, or about both.

The separated twins studied by Burt were said to be children when tested, not adults. Presumably, then, the twins were given actual (not "camouflaged") IQ tests. From a careful reading of Burt's papers, however, it is impossible to determine which, if any, IQ tests might have been given to any twins he might have studied. For documentation on this point, see Kamin, 1974.

Figures too good to be true

Furthermore, the IQ correlations that Burt claimed to have observed in his separated twins are quite literally incredible. The first reference to separated twins by Burt was in his 1943 paper. He claimed to have studied 15 pairs of separated identical twins. Their IQ correlation, on some unspecified test, was said to be 0.77. By 1955, Burt had managed to increase his sample of separated twins to 21 pairs. The level of precision in Burt's calculations had increased, and he now adopted the unusual practice of reporting his correlations to the third decimal place. The correlation was now said to be 0.771, based on a group test of intelligence. The precision of Burt's procedural descriptions had not, alas, increased.

There was no indication of which group test of intelligence might have been employed. (A group test is one which can be sat by any number of candidates at the same time, since it does not need to be individually administered.)

By 1958, Burt claimed that his sample of separated twins had been increased to "over 30". The correlation on the group test was *still* reported as 0.771—identical, to the third decimal, to that reported earlier for a smaller sample. By late 1958, Burt's research associate, Conway, was able to report that the sample of separated twins had been increased to 42 pairs. This sudden swelling of the sample size did affect the reported correlation, but not much. The correlation was now said to be 0.778. When Burt last reported on his separated twins, in 1966, the sample size was said to have increased to 53 pairs. The correlation, almost supernaturally, had returned to the originally reported 0.771.

This remarkable consistency can be observed not only in Burt's work on separated twins, but also in his work on identical twins who have been reared together, in their own families. The 1955 Burt paper claimed to have studied 83 such pairs, and to have observed an IQ correlation (on an unnamed group test) of 0.944. That correlation, it might be noted, is remarkably high. There is considerable measurement error involved in IQ testing, and it is doubtful whether if the same group IQ test were to be given on two separate occasions to the same set of people, a correlation that high would be observed between scores on the two occasions. The Burt 1958 paper, in any event, again reported a correlation of 0.944 for identical twins reared together.

The Conway 1958 paper, in remarkable synchrony with her report on separated twins, observed a trivial change in the correlation for twins reared together. It was now said to be 0.936, with the number of pairs not specified. When Burt made his final report in 1966, the correlation for twins reared together had also returned to its original value of 0.944. The sample size was said to have increased to 95 pairs.

The kinds of data collected by scientists in the real world simply do not behave with such incredible stability. When sample sizes are increased, the correlations observed will almost certainly change somewhat. Yet in Burt's work there is a repeated tendency for correlations to remain the same to the third decimal place. Thus Burt's sample of siblings reared apart increased from 131 to 151 pairs between 1955 and 1966, but correlations remained identical to the third decimal place. The Burt sample of fraternal (not identical) twins reared together mysteriously *decreased* by 45 pairs between the same two years. But no matter: correlations remained the same to the third decimal.

ATTACK AND COUNTER-ATTACK

There are many other absurdities, contradictions, evasions, ambiguities and dishonesties scattered throughout Burt's work. These were documented in detail in my earlier works (Kamin, 1973, 1974). With some measure of restraint, I wrote, after reviewing Burt's work: "The numbers left behind by Professor Burt are simply not worthy of our current scientific attention." The clear implication—that Burt had invented the data in order to support his ideas about social and educational policy—was left for the reader to make.

There are, alas, none so blind as those who will not see. Perhaps a typical reaction from the academic establishment was that of Loehlin, Lindzey and Spuhler in a 1975 work commissioned by the Social Science Research Council. They wrote: ". . . one could presumably attempt to *find out* the explanations for some of the anomalies in the data: while Burt himself is dead, doubtless some of his former students and research associates could shed light on the details of some of the researches, and it might not be out of the question to track down some of the 'unpublished theses' and 'LCC reports' that Burt refers to as the primary documentation of the studies. Kamin prefers simply to dismiss Burt's data as 'not worthy of serious scientific attention'." In England, the defence of Burt was even more succinct. The psychologist David Fulker, reviewing my critique of Burt, wrote in 1975: "Certainly, when we are told that 'the marvellous consistency of his data supporting the hereditarian position often taxes credibility', there is exaggeration."

Professor Jensen reacted more sensibly, executing what might fairly be described as a brisk about-face. Two years earlier he had extolled Burt as a born nobleman whose large and representative samples had secured his place in the history of science. But in 1974 Jensen wrote, after citing the absurdities that I had documented, that Burt's correlations and data were "useless for hypothesis testing"—that is to say, worthless. However, Jensen indicated that Burt's work had been merely careless, not fraudulent. Jensen further maintained that the dismissal of Burt's data did not substantially affect the weight of the evidence indicating a high heritability of IQ. This incredible claim was made despite Jensen's declaration, in 1969, that Burt's work was "the most satisfactory attempt" to calculate the heritability of IQ.

The *Sunday Times* exposé

The argument about Burt's data might have been confined to academic circles, and might have tiptoed around the question of Burt's fraudulence, were it not for Oliver Gillie of the London *Sunday Times*. Dr Gillie, the newspaper's medical correspondent (and incidentally also a geneticist), attempted to locate two of Burt's research associates—the Misses Conway and Howard. These two women had published papers, in collaboration with Burt and separately, in the psychological journal that Burt edited. They were the people who, according to Burt, had actually tested the twins and other relatives about whom he wrote so extensively.

There was no documentary evidence to be found, anywhere, of the existence of either of these two "research associates". Burt's colleagues at University College, London had never laid eyes on them. Nor had his secretary or housekeeper seen them, or any correspondence from them. When asked about them, Burt had sometimes maintained that they had emigrated to Australia—*before* the time when they were supposedly testing separated twins in England! Dr Gillie's front-page article, written in 1976, flatly asserted that Burt had been guilty of a major scientific fraud and cited many of the absurdities in Burt's work that were by then becoming quite widely known in academic circles. The charge of fraud against Burt was supported by the testimony of two of his distinguished former students, Alan and Ann Clarke. The cat was now out of the bag, and the feathers began to fly.

Professor Jensen wrote to *The Times* to assert that I had "spearheaded the attack . . . to wholly discredit the large body of research on the genetics of human mental abilities. The desperate scorched-earth style of criticism that we have come to know in this debate has finally gone the limit, with charges of 'fraud' and 'fakery' now that Burt is no longer here to . . . take warranted legal action against such unfounded defamation."

Professor Eysenck leapt to Burt's defence as "Britain's outstanding psychologist for many years, who had been knighted for his service to education, and who had achieved world fame for his contributions . . .". The allegations against Burt, according to Eysenck, contained "a whiff of McCarthyism, of notorious smear campaigns, and of what used to be known as character assassination". While implying that he disapproved of smear and of character assassination, Eysenck nevertheless described Dr Gillie's behaviour as "unspeakably mean". The press, according to Eysenck, had discussed the Burt affair in an irresponsible way. The tone of the press coverage had been so debased that in 1977 Eysenck threatened (he did not, alas, follow through) to retire from public debate to the privacy of his scientific garden.

This swashbuckling attack on Burt's critics was mounted before many members of the psychological community were aware of the conclusions being reached by Burt's authorised biographer, Professor Leslie Hearnshaw. With publication of Hearnshaw's work impending, the tone of Burt's defenders became more muted. Thus, by 1978, Eysenck was writing of Burt: "On at least one occasion he invented, for the purpose of quoting it in one of his articles, a thesis by one of his students never in fact written; at the time I interpreted this as a sign of forgetfulness." This lapse of memory on Burt's part had evidently been forgotten by Eysenck when, one year earlier, he had attacked Burt's critics as McCarthyite character assassins. By 1978 Eysenck was beginning to cast in his lot with the character assassins. Though Eysenck was not certain that Burt had engaged in "wholesale faking", he was now certain that Burt had behaved "in a dishonest manner".

The final blow: Burt's biography

The last lingering doubts about Burt's faking have been put to rest by Hearnshaw's painstaking biography, published in 1979. The work was commissioned by Burt's sister, and Burt's diaries, letters and papers were made available to Hearnshaw. Professor Hearnshaw had delivered the eulogy at Burt's memorial service, and he began his work as an admirer of Burt. He could find no trace of Miss Conway, or Miss Howard, or of separated twins. He found many instances of dishonesty and of evasion and of contradiction in Burt's written replies to correspondents who had asked questions about his data. The evidence was clear in indicating that Burt had collected no data at all during the last 30 years of his life, when most of the twins were supposedly studied.

With obvious reluctance, Hearnshaw was forced to conclude that the charges made by Burt's critics were "in their essentials valid", and that Burt had "fabricated figures" and "falsified". Perhaps too charitably, he suggested that Burt might actually have collected some of his purported data when he was younger, but that, as an ailing and elderly man, he padded the data and engaged in various forms of deception. From the available evidence, however, it is reasonable to suggest that perhaps Burt never tested a separated twin, or calculated a genuine correlation between relatives, in his entire life.

There is now no doubt whatever, and no dispute, that in any discussion of IQ heritability the entire body of Burt's work must be discarded. The Burt data were by far the strongest and clearest in the entire field. The following pages will document how weak and inconclusive the data from other sources are. The remaining data cannot even establish that the heritability of IQ is significantly greater than zero.

What, however, are we to make of the fact that Burt's transparently fraudulent data were accepted for so long, and so unanimously, by the "experts" in the field? When I first criticised Burt's papers, as an outsider to IQ testing, Eysenck wrote derisively, in 1974, of my "novitiate status" and my "once-a-year interest" in a subject best left to the experts. The same Burt papers that I had first read in 1972 had been read many years earlier by Eysenck, who repeatedly quoted them as gospel.

A sorry comment

Perhaps the most important moral to be drawn from the Burt affair was spelled out by NJ Mackintosh in a 1980 review of Hearnshaw's biography in the *British Journal of Psychology*:

"Ignoring the question of fraud, the fact of the matter is that the crucial evidence that his data on IQ are scientifically unacceptable does not depend on any examination of Burt's diaries or correspondence. It is to be found in the data themselves. The evidence was there . . . in 1961. It was, indeed, clear to anyone with eyes to see in 1958. But it was not seen until 1972, when Kamin first pointed to Burt's totally inadequate reporting of his data and to the impossible consistencies in his correlation coefficients. Until then the data were cited with respect bordering on reverence, as the most telling proof of the heritability of IQ. It is a sorry comment on the wider scientific community that 'numbers . . . simply not worthy of our current scientific attention' . . . should have entered nearly every psychological textbook . . . ."

To my mind, as the following pages will indicate, it is an equally sorry comment on the fraternity of IQ testers that, having lost Burt's data, they continue to assert that the remaining evidence demonstrates the high heritability of IQ.

14

SEPARATED IDENTICAL TWINS

"IQs of identical twins reared apart . . . [are] perhaps the most cogent evidence in favour of the genetic determination of intelligence. . . . If the genetic case rested on just one kind of support, this would be the one chosen by most experts."

HANS J EYSENCK, 1973

Three investigators have located a large enough number of separated twin pairs to compile their IQs statistically. All three studies reported basically similar results. Taken at face value, the results might suggest a substantial heritability of IQ. In 1937 Newman, Freeman and Holzinger in the United States found a correlation of 0.67 in 19 pairs. In 1962 Shields reported an IQ correlation of 0.77 for 37 twin pairs in England. And in 1965 Juel–Nielsen found a correlation of 0.62 in 12 Danish pairs. There are many good reasons, however, for not regarding these substantial correlations as valid estimates of the heritability of IQ. This chapter will review each of the three studies in turn, starting with the English study by Shields.

ENGLISH FINDINGS

The separated twins studied by Shields had been located by a television appeal for volunteers to co-operate in a scientific study. There is no reason to suppose that the twins represented a random sample of the population, or even a fair sample of all separated identical twins. There presumably exist in the world some identical twins who were separated at birth and who do not know of each other's existence. These are the twins who would be most likely to have experienced very dissimilar environments; but it is precisely these twins who cannot volunteer to take part in the study. Thus, inevitably, studies of separated twins are biased towards including just those pairs whose similar environments have increased their IQ resemblance.

The problem of "unequal" backgrounds

The Shields volunteers, whose ages ranged from eight to 59 years, were predominantly women, as were the subjects in the other two studies of separated twins. From the detailed case studies presented by Shields it can be ascertained that, in 27 cases, the two "separated" twins were in fact reared in related branches of the same biological family. There were only 13 pairs whose members had been reared in unrelated families. The most common pattern was for the biological mother to rear one member of the twin pair, with the other twin being reared by the maternal grandmother or an aunt.

There is clear evidence that the two sets of twins differed in many ways. To take just one example, the average age of the twins reared in related families was 42—significantly older than the average age of 32 of twins reared in unrelated families. This may reflect the fact that twins reared in related families are more likely to remain in contact with each other as they grow older and thus volunteer to be studied.

As to testing methods, the mental tests used by Shields in his study were not, unfortunately, well standardised IQ tests. To give each of his subjects a "total intelligence score", Shields lumped together the results of two different brief tests. They were the non-verbal Dominoes test, which was employed in the British army during the Second World War, and a part of Raven's Mill Hill vocabulary test. There is no satisfactory way of converting Shields' "total intelligence scores" into more orthodox standardised IQs. In examining the data, we must therefore follow Shields' method of combining "raw" (that is, unconverted) scores from the two tests to assess intelligence.

To return to the question of backgrounds, the intelligence correlation for the 27 pairs reared in related families can be calculated as 0.83—significantly higher than the correlation of 0.51 observed in the 13 pairs reared in unrelated families. This difference testifies to the importance of environment in determining how closely the IQs of "separated" identical twins will resemble each other. Though in every case the twins shared identical heredity, it is those pairs reared by relatives—and thus experiencing similar environments—who were strikingly alike in IQ. The fact that pairs reared in unrelated families nevertheless correlated 0.51 in intelligence must not be taken as unambiguous evidence for the role of heredity, because even among such pairs it was common for one twin to be reared by the mother and the other by close family friends. None of Shields' twins, then, can be said to have been reared in very different social conditions.

These figures can be broken down further. There were 24 cases in which the mother herself reared one of the separated twins. In 12 of these cases, the remaining twin was placed with a relative of the mother. In the other 12, the remaining twin was placed elsewhere, sometimes with a relative of the father. The intelligence correlation for the first group (in which the outside twin was reared by a maternal relative) was a striking 0.94—very significantly higher than the 0.56 in the second group. Where the outside twin was with a paternal relative, Shields' case studies sometimes reveal severe "in-law" troubles: these included solicitors' letters about custody, refusals to speak to each other, hostility towards the mother-in-law, and face-slapping episodes. When one twin was reared by the mother and the other by her relative, they presumably experienced more similar environments and had more contact with each other. The correlation between such twins, in fact, is as large as any that has been reported for identical twins who have not been separated at all!

There is reason to doubt whether many of the twins studied by Shields—or by the other researchers—in fact experienced much separation. To be counted as a separated twin pair by Shields it was only necessary that, at some time in childhood, the two children had been reared in different homes for at least five years. This meant that some of Shields' pairs had not been separated at all until the age of seven, eight or nine. The following examples from Shields' case histories show just how similar were the environments experienced by most of the pairs he studied.

Some of Shields' case studies

Jessie and Winifred were separated at three months. "Brought up within a few hundred yards of one another. . . . told they were twins after the girls discovered it for themselves, having gravitated to one another at school at the age of five. . . . They play together quite a lot . . . . Jessie often goes to tea with Winifred . . . . They were never apart, wanted to sit at the same desk . . . ." There is considerable unconscious humour here. The investigator who has provided us with more than half the documented cases of "separated" identical twins here informs us that a "separated" pair of eight-year-olds "were never apart"! These twins, it might be noted, were reared by *unrelated* families. A twin pair reared by related families might have even more contact.

We might also consider Bertram and Christopher, said to have been separated at birth: "The paternal aunts decided to take one twin each and they have brought them up amicably, living next door to one another in the same Midlands colliery village . . . They are constantly in and out of each other's houses." Or take Odette and Fanny, who from the ages of three to eight exchanged places every six months—one going to the mother, the other to the maternal grandmother. Or Benjamin and Ronald, "brought up in the same fruit-growing village, Ben by the parents, Ron by the grandmother . . . . They were at school together . . . . They have continued to live in the same village . . ." until, at the age of 52, they were tested by Shields. Or, lastly, Joanna and Isabel, aged 50, who had been "separated from birth to five years" but then "went to private schools together".

The study of separated identical twins, remember, would be of unique value if it could be assumed that there was no similarity between the environments in which they were reared. Professor Burt, who provided

Children from different races and backgrounds sharing the same classroom environment. The experiences and genetic make-up they share are not enough to help us investigate the determinants of IQ.

no case studies, was able to report that there was no such correlation. The case studies of real twins provided by Shields show clearly that, in the real world, the environments of so-called "separated" twins are massively correlated. It is not therefore necessary to attribute the observed IQ correlation to heredity. It might be largely, or entirely, due to the highly similar environments.

The problem of unconscious bias

There are still other reasons why the reported IQ correlation of separated twins should not be attributed solely to their genetic identity. The scrupulously detailed Shields case studies specify that, in the case of 35 pairs, Shields himself tested both twins. With the remaining five pairs, the twins were tested by *different* examiners. We can calculate the intelligence correlation of these two categories. Where both twins were tested by Shields, it works out at 0.84, compared with the trivial 0.11 for pairs where a different examiner tested each twin. Despite the very small size of one of the samples, the two correlations differ to a statistically significant degree. This suggests the possibility that *unconscious bias* on the part of the tester may have inflated the IQ correlation of separated twins.

This suggestion should be clearly understood. There is no implication that Shields was in any way dishonest; indeed, his detailed case studies are a model of scientific explicitness and integrity. The fact is, however, that the theories and wishes of experimenters often influence the behaviour of their subjects, a fact long recognised in experimental psychology. That is why experimenters are often kept "blind" about what the subject with whom they are working, whether animal or human, is "supposed" to do. Yet this precaution was not taken in the Shields, or any other, study.

That the behaviour of the person administering an intelligence test may affect the person taking it is obvious. The "non-verbal" Dominoes test used by Shields, for example, requires the tester to give complicated instructions to, and work out specimen examples with, the test-taker. There would be nothing surprising if, entirely unconsciously, the tester were to give highly similar instructions, encouragements or discouragements to each member of a twin pair.

When one considers what a precious scientific resource separated identical twins are supposed to be, it is astonishing that in *all* studies of such twins the same tester routinely gave the test to both members of the pair. There is no question that a preferable procedure would be for a different examiner to test each of the two twins, with neither examiner knowing the other's results. That way, the tester's theoretical expectations could not unconsciously bias the administration or scoring of the test.

The suggestion that unconscious bias might have inflated Shields' intelligence correlation has been vigorously resisted by hereditarians. Thus Fulker has sought to explain the large discrepancy in intelligence scores among the five pairs tested by different examiners. Some of these pairs, he pointed out in 1975, had been widely separated geographically and might have dissimilar IQs for *that* reason. This is true enough—though it is an environmental argument rather than a testimonial to the overwhelming power of the genes to guarantee similar IQs in identical twins. And in one case, one of the twins had a history of congenital syphilis, amnesia and recurring blindness which, Fulker suggested, was quite sufficient to explain the large score difference. What neither Fulker nor those who quote his critique point out is that it was the blind, amnesiac and syphilitic twin who had by far the higher IQ.

The problem of "invalid" scores

Finally, it should be pointed out that in my analysis of Shields' data I have included all 40 pairs to whom tests were administered—including three pairs discarded by Shields, on the grounds that their tests were "invalid". One pair was excluded, for example, because Shields felt that the very low Dominoes score of one twin meant that she had failed to understand the instructions; this twin, of course, had a much lower total intelligence score than her sister. The twins had been tested by different examiners. When Shields later had them retested on Wechsler's individual IQ test, again by different examiners, their IQs were found to be 92 and 111. This 19-point IQ difference is one of the largest ever observed in a pair of separated twins; the only larger difference is one of 24 points in the Newman study. It is clear that the twin who "didn't understand" the Dominoes instructions had a much lower IQ than her sister, and the pair can therefore validly be included in the analysis of Shields' data.

AMERICAN FINDINGS

The American study of 19 separated pairs by Newman, Freeman and

Holzinger (which, for convenience, we shall refer to as the Newman study) shares all the weaknesses of the Shields report. Again there was an obvious tendency to include only those twins who resembled one another strongly. The volunteers, responding to newspaper and radio appeals, had to be brought to Chicago for study, often at considerable expense. The study was done during the Great Depression, and the researchers could not afford to transport to Chicago volunteers who, on medical examination, might turn out not to be identical. They therefore mailed a questionnaire to all volunteers, who had to attest that they were "strikingly similar" and to send photographs. When a pair who looked so alike that they were mistaken for each other wrote that they were "as different as can be in disposition", they were excluded from the sample! Only those who described themselves as being very much alike were accepted. With this kind of biased selection of subjects, it is perhaps surprising that the IQ correlation found by Newman was as low as 0.67.

There was, as in the Shields study, an obvious similarity in the environments in which the "separated" twins were reared. Thus, Kenneth and Jerry were adopted by two different families. Kenneth's foster father was "a city fireman with a very limited education". Jerry's foster father was "a city fireman with only fourth-grade education". The two boys lived between the ages of five and seven in the same town (where their fathers were firemen) but "were unaware of the fact". Likewise Harold and Holden were said to be "separated", but each was adopted by a family relative, they lived three miles apart, and they attended the same school.

The problem of poor standardisation

The IQ test used by Newman was the 1916 version of the Stanford–Binet, which contained a form for adults but was designed primarily for children. Standardisation of the 1916 Stanford–Binet, even for school-children, had been notoriously poor. IQ tests should in theory provide an average IQ of 100 at every age, but scores on the 1916 Stanford–Binet were negatively correlated with age. In other words, the older a child became, the less intelligent the test declared him or her to be. Standardisation of the adult questions had been even less thorough. The sample of adults used to standardise the test was quite inadequate; indeed, it contained no women. Yet most of the twins in the Newman study were adult women.

The defective standardisation of an IQ test poses a very serious problem for the study of separated identical twins. When a test is not perfectly standardised—and no test is—one sex or the other will tend to receive higher IQ scores; and people will tend to receive lower (or higher) scores according to their age when tested. Identical twins are necessarily of the same age and sex. Thus, to the degree that people of the same age and sex receive similar scores on a given test, the IQ correlation of identical twins will be artificially inflated. Part of the similarity in twins' test scores must be due to their being of the same age and sex—not to their identical heredity. There is evidence (see Kamin, 1974) that in the Newman study the observed IQ correlation between separated twins is at least partly an artifact of the very poor standardisation of the 1916 Stanford–Binet.

The problem of volunteer subjects

Reading the Newman report, one is forced to recognise that in studies of this sort the researchers depend heavily on the accuracy of volunteers' information. At one point in their book, Newman and his colleagues state that Ed and Fred had "lived without knowledge of each other's existence" for their entire 25 years. Both were said to have worked as electrical repair men for the telephone company, and to have owned fox terriers named Trixie. The case study tells a different story. "[The twins] went to the same school for a time, but never knew that they were twin brothers. They had even noticed the remarkable resemblance between them, but they were not close companions. When the twins were about eight years old, their families were permanently separated . . . . There is evidence that Edwin had more continuous and better instruction, though the actual facts are difficult to obtain."

The statement that Ed and Fred attended the same school before their families separated simply does not square with the assertion that the two had lived their lives not knowing of each other's existence. Perhaps accounts of identical jobs, and fox terriers named Trixie, should be regarded with scepticism in a case where "actual facts" about such straightforward matters as education and separation are "difficult to obtain". The twins could scarcely be blamed if, in a misguided effort to co-operate with science, or to inject romance into their life story, they stretched a fact or two. There is, in most scientific work, a sharp distinction drawn between evidence and anecdote. The boundary here appears to be blurred.

DANISH FINDINGS

The Juel–Nielsen Study, involving a mere 12 pairs of Danish separated twins, adds little to the picture. Using a Danish translation of Wechsler's test for adults, Juel–Nielsen reported a correlation of 0.62. There had never, however, been any Danish standardisation of Wechsler's test. The males in Juel–Nielsen's sample had significantly higher IQs than the females and IQs appeared to vary considerably with age, which artificially inflates the reported correlation, of course.

The problem of negative correlations with age

The Juel–Nielsen twins were also tested with Raven's Progressive Matrices, a "non-verbal" test in some ways similar to the Dominoes test used by Shields. The analysis of this test, like Shields' analyses, must be based on raw scores rather than on IQs. The correlation between twins' scores on this test was accurately reported as 0.77 by Eysenck in 1979. The Eysenck account neglects to inform the reader, however, that twins' ages and their scores on Raven's test produced a robust *negative* correlation of −0.65. Thus in the Juel–Nielsen study, older twins did worse on the Raven test than younger ones. This effect of age on test score served to inflate considerably the observed correlation of identical twins.

We might by now suspect that the environments of Juel–Nielsen's "separated" twins had been highly similar—and indeed they were. Thus Ingegard and Monika were cared for by relatives until the age of seven, then lived with their mother until they were 14. "They were usually dressed alike and very often confused by strangers, at school, and sometimes also by their step-father. . . . The twins always kept together when children, they played only with each other and were treated as a unit by their environment . . . ." Twins such as these, remember, are described by hereditarian theorists as "separated". The unsuspecting student who reads in a textbook that the IQs of separated identical twins are highly correlated is not likely to conjure up an image remotely resembling the reality of Ingegard and Monika and will believe that science has shown IQ to be highly heritable.

AN OVERVIEW

Taken as a whole, the studies of separated identical twins provide no unambiguous evidence for the heritability of IQ. The apparently most impressive study has been unmasked as a fraud. The most obvious defect of the remaining three studies is the glaring tendency for the environments of so-called separated twins to be highly correlated. This tendency, no less than identical genes, might easily be responsible for the observed resemblance in IQs. We cannot guess what the IQ correlation would be if, in a science fiction experiment, we separated pairs of identical twins at birth and scattered them *at random* across the full range of available environments. It could conceivably be zero—which would force us to conclude that the heritability of IQ is zero.

Similarities of environments apart, we have noted that IQ correlations are artificially raised by excluding from the samples those pairs whose experiences have been most dissimilar, and by defective standardisation of tests for sex and for age. We have also noted evidence to suggest that correlations may have been inflated by unconscious tester bias; certainly, no study has taken the precaution of eliminating such bias.

Professor Eysenck has said that if the case for the heritability of IQ is to stand or fall on one kind of evidence, he and other experts would bank their all on the study of separated identical twins. There are, however, other, less solid, forms of evidence put forward by hereditarians, which we shall examine in the following pages.

15
STUDIES OF ADOPTED CHILDREN

"Usually foster children are employed . . . in order to avoid the contamination of environmental with genetic factors. It is essential, of course, that the adoption agency should not be placing the children selectively."

HANS J EYSENCK, 1971

The fact that parents and children resemble each other in IQ does not in itself tell us anything about the relative importance of heredity and environment. The problem, of course, is that parents provide their children both with genes and with environment. The high-IQ parent is likely to provide his or her child with intellectual stimulation in the home, and is likely to stress the importance of doing good school work. The same parent has transmitted his or her genes to the child. There is no way, in ordinary families, of separating the effects of genes from those of environment. The great virtue of studies of adoptive families is that, in theory at least, they allow us to separate genetic from environmental transmission. The adoptive parent provides his or her child with environment but not with genes. Thus the IQ correlation between adopted child and adoptive parent is of considerable theoretical interest—particularly when it is compared to other relevant IQ correlations.

THE MEASUREMENT OF SES

There are several forms of adoption study, but none has provided an apparently more clear-cut result than the 1975 report by Munsinger. The Munsinger study was based on 41 adopted children in California, for whom IQ scores were available; all had been separated from their biological parents in earliest infancy and reared by adoptive parents. Though there were no IQ scores available for either the biological or the adoptive parents, Munsinger was able to obtain an individual rating of socio-economic status (SES) for each parent. Munsinger correlated the child's IQ with the SES of (a) its adoptive parents and (b) its biological parents. From each set of parents, adoptive and biological, the average SES of the two mates—the "midparent SES"—was calculated.

The Munsinger report showed no relation at all between the child's IQ and the SES of its adoptive parents. The correlation between a child's IQ and the adoptive midparent SES was in fact -0.14: although not statistically significant, there was a slight tendency for the adopted children of high-SES adoptive parents to have lower IQs. But the correlation between a child's IQ and the biological midparent SES was an astonishing 0.70—higher even than the one normally observed in ordinary families where children are reared by their biological parents. Taken at face value, the Munsinger results imply that upper-class parents have genes for high IQ, that the child who receives those genes from his biological parents will develop a high IQ even if he never sees them, but that the child adopted by upper-class parents will not benefit from their genes or from the superior environment they provide. IQ, it would seem, is determined exclusively by biological inheritance.

Were Munsinger's ratings accurate?

There are many reasons, though, for not taking Munsinger's results at face value. The original paper provided almost no information about how the ratings of parental SES had been arrived at and therefore did not rule out the possibility that they might have been biased—quite possible if the person making the ratings had knowledge of the child's IQ while rating parental SES, for example. When I raised this possibility with Munsinger in private correspondence, he ruled it out categorically. He wrote in a letter of November 6, 1975: "If you mean the translation from occupation to numbers, then the reliability is over 0.98 based on two different blind judges . . . . All the ratings of SES were done by two people independently, and with no knowledge of the child's IQ."

The tables of raw data published by Munsinger provide the SES ratings, on a six-point scale, for each individual biological and adoptive parent. There are thus SES ratings given for 82 couples. For 48 of those couples, the SES of the two mates is identical. That makes perfectly good sense: it is well known that people tend to select mates from their own social class. The incredible fact, however, is that in all 34 cases where SES is not identical, the partners differ by precisely *two* social classes—never by one, or by three, four, or five, but always and precisely by two! Such a strange reluctance to mate with members of an adjacent social class—while succumbing to the charms of individuals precisely two social classes removed—is clearly nonsensical. There is painfully obvious error of some kind in Munsinger's SES ratings.

This absurdity was something I drew attention to in 1977 in a critical comment published in the same journal that carried Munsinger's original report. The journal also published a reply to my criticism, in which Munsinger wrote that he could not "report precisely" how the ratings had been "generated". A new attempt to rate the occupations of the same parents on the SES scale had indicated that rating was "a difficult, subjective, and sometimes ambiguous procedure". These belated comments simply do not square with Munsinger's earlier private assurance to me that the original ratings had been done blindly by two different judges, with a reliability over 0.98.

Cause for rejection

The Munsinger data, like those of Cyril Burt, appear too good to be correct. There is some irony in the fact that a leading hereditarian writer, Herrnstein, explicitly pointed to Munsinger's report as a worthy replacement for Burt's discredited studies. The Munsinger data must now be discarded, along with Burt's. We might also reflect on the fact that work containing such an obvious error can be published in a leading journal of behavioural genetics. The critical standards in this field do not seem to have improved dramatically since Burt's day.

THE CLASSIC DESIGN

The classic studies of adoption carried out by Burks in 1928 and by Leahy in 1935 were designed with a different logic (see Figure 26). Instead of obtaining information about the biological parents of adopted children, Burks and Leahy obtained the IQ scores of the children and their adoptive parents and calculated the correlation. This correlation, which was taken to reflect the effect of environment alone, could then be compared to the correlation between biological parent and child in a "matched control group" from ordinary families. In the control group, the parent–child IQ correlation should have reflected the effects of environment *plus* genes and should, if genes are important determiners of IQ, have been much higher than in the adoptive families. In both studies it was. The average parent–child correlation reported by Burks

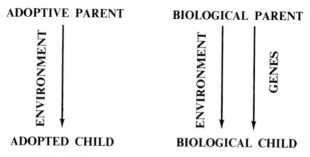

Fig. 26. The "classical" adoption design of Burks (1928) and of Leahy (1935). Note that correlations in two different, but supposedly matched, groups of families are compared. In the biological families, parent transmits environment *plus* genes to child.

and Leahy was 0.48 in the biological families, compared to only 0.15 in the adoptive families, suggesting that environment plays quite a small part and genes a large one.

Restricted variance in adoptive families

This comparison makes sense, however, only if we are convinced that the biological families used as a control group were "matched" to the adoptive families in a meaningful way. There are many characteristics of adoptive families which might depress the magnitude of the correlation between adoptive parent and adopted child. To begin with, all adoptive parents—though not necessarily all control parents—actively wanted a child. The adoptive parents had all been selected by adoption agencies as especially fit parents—economically secure, emotionally stable, not alcoholic, without a criminal record, and so on. It seems very likely, then, that *all* adoptive families would provide much better than average environments for their children, and that adoptive parents would all tend to have quite high IQ scores. The necessary statistical consequence of such *restricted variance* is that the parent–child IQ correlation in adoptive families cannot be very high—even if IQ variation is determined by environment.

To understand this technical point, consider the correlation between a man's weight and his success as a boxer. It would be very high if boxers of all weights were allowed to fight each other because the heavyweight boxer would almost always defeat the lightweight. To avoid such a correlation, definite weight divisions have been established by boxing authorities. Fights can only take place between boxers of reasonably similar weight, and the correlation between weight and boxing success is consequently very low. We are suggesting that in terms of the environments provided for their children almost all adoptive parents—unlike biological parents—are in the heavyweight division. That would account for the lower parent–child IQ correlation observed in adoptive families. The correlation would presumably be much higher if parents who would provide poor environments wanted to, and were allowed to, adopt more often.

Less-than-perfect match

Of course, both Burks and Leahy attempted to "match" their biological families to their adoptive families in at least some ways. The children in the biological families were matched to the adopted children for both age and sex, so that the adoptive parents—most of whom had tried to have their own children—were significantly older than the biological parents. For obvious reasons, there were fewer siblings in the homes of the adopted children. The two groups of parents were matched for occupational level, for years of education and for "type of neighbourhood". Despite this matching, the income of the adoptive parents was 50 per cent higher than that of the biological parents, and their homes were 50 per cent more expensive. This makes it clear that adoptive and biological families cannot meaningfully be regarded as "matched" simply because they are comparable on a few rough measures of occupation, education or whatever. Couples who want to and are allowed to adopt are obviously a very special group, and their special successful characteristics are not adequately captured by demographic measures of "environment". There is considerable evidence in the Burks and Leahy studies (see Kamin, 1974) to indicate that the environments of adoptive families are not only richer than those of the "matched" biological families but also more restricted in variance. All this nullifies the validity of comparing the parent–child IQ correlations of the two types of family.

THE NEW IMPROVED DESIGN

There is, however, an obvious improvement on the classical Burks–Leahy design—an improvement which avoids the impossible requirement of matching adoptive and biological families (see Figure 27). There are many adoptive parents who, in addition to adopting a child, have a biological child of their own. The new design correlates a parent's IQ with the IQ of (a) the adopted child and (b) the biological child. The two children have been reared in the same household by the same parent, but to the extent that genes determine IQ, the correlation between parent and biological child should obviously be larger. The parents in all such families have, of course, been selected by adoption agencies. We can therefore expect restricted environmental variance and relatively low correlations. That should be true, however, for both adopted and biological children, for we are now dealing with a single group of families.

Two recent studies have employed the suggested new design—the 1977 study of Scarr and Weinberg in Minnesota and the 1979 one of

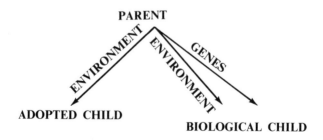

Fig. 27. The new adoption design of Scarr and Weinberg (1977) and of Horn et al. (1979). Note that only one set of families is involved, with each family containing both an adopted and a biological child. The parent transmits environment *plus* genes to the biological child.

Horn, Loehlin and Willerman in Texas. Interestingly, both studies were performed by eminent behaviour geneticists, who could scarcely be described as rabid environmentalists, and who clearly expected to discover evidence supporting a high heritability of IQ.

What counts is the mother, not the genes

The results for mothers are presented in Table 3. A mother's IQ, remember, has been correlated with the IQ of her adopted and that of her biological child. There is no significant difference in the two correlations; in Texas the mother is a trifle more highly correlated with her adopted child, in Minnesota with her biological child. The Minnesota study, it is worth noting, was based almost entirely on cases of transracial adoption. That is, the mother and her biological child were both white, and her adopted child was black. The adopted black and biological white child resembled the mother equally in IQ. The results from Texas and Minnesota appear to inflict fatal damage on the notion that IQ is highly heritable, for they show that children reared by the same mother resemble her in IQ to the same degree, whether or not they share her genes.

Table 3. Mother-child IQ correlations in adoptive families containing biological children.

	TEXAS STUDY	MINNESOTA STUDY
Mother × Biological Child	0.20 (N = 162)	0.34 (N = 100)
Mother × Adopted Child	0.22 (N = 151)	0.29 (N = 66)

("N" refers to the number of mother-child pairings on which each tabled correlation is based.)
Texas study is Horn et al., 1979; Minnesota study is Scarr and Weinberg, 1977.

Table 4. Father-child IQ correlations in adoptive families containing biological children.

	TEXAS STUDY	MINNESOTA STUDY
Father × Biological Child	0.28 (N = 163)	0.34 (N = 102)
Father × Adopted Child	0.12 (N = 152)	0.07 (N = 67)

("N" refers to the number of father-child pairings on which each tabled correlation is based.)
Texas study is Horn et al., 1979; Minnesota study is Scarr and Weinberg, 1977.

The results for fathers, presented in Table 4, appear more consistent with the idea that IQ might be heritable, particularly the Minnesota results. A number of after-the-fact explanations might be offered for the apparent discrepancy, but in the absence of more data none would be convincing. There is, however, a further bit of relevant information made available by Professor Scarr in personal communication. The educational levels of the adoptive parents in the Minnesota study were available and were correlated with the children's IQs. The education of the parents—both father and mother—correlated significantly with the IQ, of both the adopted and the biological child, slightly more so with the former, which suggests that, with fathers as with mothers, genes are not very important in determining resemblance between parent and child.

Siblings will be siblings, whatever their origin

The families represented in Tables 3 and 4 also give rise to three types of sibling relationship, some genetic, some not. There are, firstly, biologically related pairs of siblings (biological children of the adoptive parents); secondly, biologically unrelated pairs of adoptive siblings (both children adopted by the same family); and finally, biologically unrelated pairs (one biological child and one adopted child of the same parents). The IQ correlations for the three types are presented in Table 5. The

Table 5. Sibling IQ correlations in adoptive families containing biological children.

	TEXAS STUDY	MINNESOTA STUDY
Biological–Biological Pairs	0.35 (N = 46)	0.37 (N = 75)
Adopted–Adopted Pairs	———	0.49 (N = 21)
Biological–Adopted Pairs	0.29 (N = 197)	0.30 (N = 134)

("N" refers to the number of sibling pairings on which each tabled correlation is based.)

Note that biological-biological pairs are genetically related, and that other two types of pairs are not. Texas study is Horn et al., 1973; Minnesota study is Scarr and Weinberg, 1977.

results are clear. None of the correlations differs significantly from any other, and there is no indication that the correlation for genetically related siblings is higher—another fatal blow to the view that IQ is highly heritable.

Whether or not they share common genes, then, two children reared in the same household resemble one another to the same degree. This should not surprise us, for, though behaviour geneticists have tended to ignore the finding, Freeman, Holzinger and Mitchell reported as long ago as 1928 that the IQs of adopted children correlated just as highly with measures of home environment as did the IQs of biological children reared in the same homes.

AVERAGE IQs

Up till now we have examined only *IQ correlations* in adoption studies, not *average IQs*. Yet the average, or mean, IQ of adopted children is of considerable interest. The authors of the Texas adoption study, for example, concluded that, overall, their correlations indicated "moderate heritabilities" for IQ. But the analysis of average IQ levels, much to their surprise, "suggests a heritability of IQ that is close to zero". That is a surprising suggestion to issue from the pen of behaviour geneticists and it merits close examination.

The Texas story

The Texas investigators were able to obtain IQ scores for the biological mothers of children surrendered for adoption. They had IQs significantly lower, by about six points, than the adoptive mothers who reared their children. But the adopted children had an average IQ every bit as high as the adoptive mothers' biological children—about 112 in each case, a very high IQ. These figures indicate that adopting parents successfully transmitted high IQs to all the children they reared, whether or not they shared genes with them. The relatively low IQs of adopted children's biological mothers simply did not matter. The average IQ of the adoptive parents was about 114, and there was very little variation among them (in technical terms, the standard deviation of their IQs from the mean was about 11, as against 15 in the general population), or among their children, whether biological or adopted.

The Minnesota story

The adopted children in Scarr and Weinberg's transracial adoption study also had very high average IQs. They were calculated by me from the raw data generously made available by Professor Scarr. The 56 children given the Stanford-Binet test and placed at under one year of age with couples who also had a biological child of their own had an average IQ on the Stanford–Binet of 109. This was high, but significantly lower (by 6.6 points) than the IQ of the 32 biological children in the same families. These results appear to contradict those of the Texas study. Though the adoptive parents in the transracial adoption study have endowed their adopted children with higher-than-average IQs, their biological children appear to have even higher IQs.

Age at adoption—the key to the difference?

The difference between the two studies, however, can plausibly be attributed to the fact that in Texas all the infants were placed in their adoptive homes directly from the hospital, whereas the Scarr and Weinberg study included children adopted as much as a year after birth. Three-quarters of the children had been placed between birth and eight months, and their average IQ was 111. The remaining one-quarter, placed between eight months and 12 months, had a significantly lower IQ of 103. The sooner a child is placed into an adoptive home, then, the higher its IQ is likely to be. It seems probable that if the Scarr and Weinberg children had been placed at birth, their IQs would have been equal to those of their adopted siblings, as in the Texas study.

THE EFFECT OF SELECTIVE PLACEMENT

There is still another relevant finding from adoption studies. When some academic measure is available from an unmarried mother—either her IQ or her educational level—it is often found to be significantly correlated with the IQ of the child she has had adopted, even if she has never lived with the child. The observed correlation varies, according to the study. In 1938 Snygg reported a correlation of only 0.12, while Skodak and Skeels, using the same test, reported a correlation of 0.44 in 1949. Without exception, all the studies have found some correlation, which suggests to hereditarians that the unmarried mother has transmitted IQ-influencing genes to her child.

There is an obvious alternative interpretation, however. When adoption agencies place a child, they try to fit the child to the home. Agency workers probably believe that IQ is largely genetic and that "bright" children should be placed in "good" homes. The agency may know the IQ of the unmarried mother and may even have tested her. It will certainly know her educational level, and perhaps that of her mate. It will have investigated in detail the homes of potential adoptive parents. There is therefore considerable scope for *selective placement*, with children of highly educated and high-IQ mothers going to superior homes which foster high IQ. Such selective placement, the possibility of which must always be borne in mind in interpreting the results of adoption studies, could establish a *non-genetic* correlation between the IQ of an unmarried mother and that of the child she has had adopted.

There is evidence (see Kamin, 1974) to show that selective placement is routinely practised by adoption agencies. But can it reasonably account for the observed correlations between an unmarried mother and the child she has had adopted? Thanks to the Texas and Minnesota studies, we can test this possibility by calculating the correlation between the unmarried mother and the biological child of the couple who adopted her child. Since there is no genetic relation between them, any correlation must be the result of selective placement, and of nothing else. If it is lower than her correlation with the child she put out for adoption, a genuine genetic effect might be indicated.

Unmarried mothers: does their brightness count?

Scarr and Weinberg reported that an unmarried mother's education showed a correlation of 0.32 with the IQ of the child she had had adopted. The correlation with the IQ of her child's adoptive sibling was 0.15—significant, but lower. The difference led Scarr and Weinberg to conclude that a genetic effect over and above selective placement had been demonstrated. But the vast majority of the adopted children in their study were tested with the Stanford–Binet test, while most of the biological children were tested with the Wechsler test—and the two tests can give substantially different results. (Two different tests were used because the children varied considerably in age: the adopted children were on average younger than the biological children and were thus more often tested with the Stanford–Binet.)

On the basis of the raw data of Professor Scarr, and using only Stanford–Binet scores, I have calculated new correlations for these same relationships. For 79 pairs of unmarried mother and relinquished child, the correlation was 0.28. For 29 pairs of unmarried mother and her child's adoptive sibling, the correlation was 0.33—despite the absence of any genetic relation. Thus, when the type of test is held constant, the entire correlation between an unmarried mother's educational level and the IQ of her relinquished child seems (at least in this instance) to be due to selective placement.

The same type of comparison can be made with the Texas data, but in this case the correlations involve the unmarried mother's IQ rather than her educational level. The unmarried mother is correlated 0.31 with her relinquished child, and only 0.08 with her child's adoptive sibling. These results are rather different from those observed in Minnesota; but then although they were given the same tests, the adopted children in Texas were significantly younger than their adoptive siblings. In Texas, the unmarried mothers correlated 0.19 with *other* adopted children (not their offspring) reared in the same home as their child. This significant correlation can only reflect selective placement.

The least that can be said is that selective placement accounts for a considerable portion of the correlation between unmarried mothers and their relinquished offspring. Possibly, with a sufficiently fine-grained analysis of the raw data, it might account for all of it. But selective placement could operate in either of two ways. We have stressed that the "better" adoptive homes in which children of high-IQ unmarried mothers are placed provide excellent environments—and that those environments elevate the IQs of both adopted and biological children. The determined hereditarian might argue that biological children reared in those homes have high IQs because their parents transmitted superior genes to them—and that the adopted children in such homes have high IQs because *their* biological parents also had superior genes. To argue in this way, however, one would have to assume an extraordinarily efficient selective placement process—efficient enough to result in adopted and biological children in the same family having identical average IQs.

ADDITIONAL FINDINGS

For the sake of completeness, we should note the results of another Minnesota study, also conducted by Scarr and Weinberg, this time in 1978. The adopted children in this study were adolescents when they were tested for IQ, and all were white. The design of the study, unfortunately, was of the old-fashioned Burks–Leahy variety: there were two entirely separate groups of families—a group of adoptive families, and a rather casually assembled group of volunteer biological families.

IQ correlations within the adoptive families, Scarr and Weinberg stressed, were very low. There was, in fact, no correlation at all between the IQs of two unrelated children adopted by the same parents. This highly unusual finding is obviously related to the severe restriction of variance in this study. (In statistical language, the standard deviation of the adopted children's IQs from the mean was a stunningly low 8.95, whereas in the general population it is 15.) The mothers who had given up children for adoption were believed to have had a normal average IQ, and a normal variation from the mean. Therefore the decreased variance in IQ among the adopted children is not a genetic effect, but is attributable to the similar "heavyweight" environments provided by adoptive parents. A look at the IQ scores of pairs of unrelated adopted siblings (two biologically unrelated children adopted by the same family) makes this clear: the average difference in their IQ scores was in fact a mere 11.75 points—close to the usual 12-point IQ difference observed among biological siblings. The Scarr and Weinberg findings demonstrate, then, that adolescent adopted siblings are very much alike in IQ, although, because of severely restricted variance, the effect cannot be observed in the correlation coefficient.

A big boost for IQ

The last adoption study to be discussed was conducted in France by Schiff and co-workers, who reported in 1978, and contains a number of special features. The investigators managed to locate 32 children born to lower working-class parents but adopted by high-SES parents at less than six months of age. They also obtained data for 20 *biological* siblings of the adopted children; these biological siblings had been reared by their own mothers. Thus the two groups of siblings are genetically equivalent, but one group has been reared by upper-SES (adoptive) parents, and the others by lower-SES (biological) parents. The adopted children had an average IQ of 111—a full 16 points higher than that of their stay-at-home siblings. Perhaps more important, fully 56 per cent of the stay-at-homes had failed a year in the French school system, compared to only 13 per cent of the adopted children. The title of the Jensen article which spurred the renewed interest in IQ heritability was "How much can we boost IQ and scholastic achievement?" The Schiff study gives an unequivocal answer as to what could be done if low-SES children were to be reared in richer environments.

NO CONVINCING CASE

The reader might reasonably suspect that in summarising the results of adoption studies I have selectively stressed those aspects of a complex set of data that minimise the importance of heritability. It is therefore interesting to note the conclusions reached by Professor Loehlin, one of the authors of the Texas report. While admitting that the data on average levels of IQ suggested a heritability of zero, Loehlin later went on to fit the correlations to a complicated model known as a "path model". This entails making a number of implausible simplifying assumptions, the effect of which is to increase the estimate of heritability. Even with such a heredity-loaded model in hand, Loehlin reported that the Texas correlations suggested a heritability of only about 38 per cent—a far cry from the 80 per cent figure so confidently put about by authorities such as Eysenck and Jensen. The 38 per cent figure is a little closer to zero than it is to 80 per cent. With improved experimental designs and more refined data analyses, the heritability estimates derived from adoption studies can be expected to move even closer to zero.

This review of adoption studies, like the review in the last chapter of separated twin studies, has failed to yield convincing evidence for the heritability of IQ. Though early studies appeared to suggest a high heritability, they ignored the restricted environmental variance of adoptive families. They also ignored the profound effects of selective placement. With improved designs and increased sophistication of analysis, the more recent studies of adoption produce a radically lower estimate of heritability. In fact the possibility cannot be excluded that IQ heritability is actually zero. Ten years after the publication of Jensen's article, even behaviour geneticists who conduct adoption studies have begun to point out that some of the data do indeed suggest zero heritability. The importance of the new adoption data is difficult to exaggerate. A fundamental re-evaluation of earlier research is under way.

A caring home in which reading skills are encouraged and rewarded.

16
MZ AND DZ TWINS

"There is not a shred of evidence to suggest any special differential treatment of MZ twins relevant to cognitive development. Indeed, what evidence we have is entirely negative."

HANS J EYSENCK, 1979

The most common type of study aimed at demonstrating the heritability of IQ involves comparing the two fundamentally different kinds of twins—monozygotic (MZ) or identical twins and dizygotic (DZ) or fraternal twins. The rarer MZs are the result of the fertilisation of a single ovum by a single sperm. There is an extra split of the zygote early in development resulting in the mother bearing two separate individuals. The members of a pair of MZ twins are the only human individuals whose genes are literally identical. They are always of the same sex, and typically—but not always—they are strikingly similar in physical appearance.

The more common DZs result when two separate sperms fertilise two separate ova at about the same time. The mother bears two individuals, but the two are no more alike genetically than are ordinary brothers and/or sisters. They are, indeed, ordinary siblings who happen to be conceived and born at the same time, and, like ordinary siblings, they will share, on average, about 50 per cent of their genes. They may be of the same or different sexes, and their physical resemblance is about the same as that of ordinary siblings.

IDENTICAL TWINS

If a trait like IQ is genetically determined, one would expect MZ twins to be very highly correlated in IQ—if heritability is very high, almost perfectly correlated. The IQ correlation expected among DZ twins is obviously much lower—under the simplest genetic model, only half as high as the correlation among MZs. There have been literally dozens of studies comparing the IQ correlations of MZ and DZ twins. (To rule out complications arising from possible sex differences, the DZ samples usually consist of pairs of the same sex only.) The results of these studies, almost without exception, demonstrate that the IQ correlation of MZs is considerably higher than that of DZs. Typically, the correlations reported for MZs range between about 0.70 and 0.90, compared with between about 0.50 and 0.70 for same-sex DZs. There is no doubt at all that, empirically, the MZ correlation is higher.

Hereditarians attribute this difference to the greater genetic similarity of MZ twins. There are, however, obvious environmental reasons to expect higher correlations among MZ than among DZ twins. The environments to which members of an MZ pair are exposed tend to be strikingly alike—perhaps more so than those of any other individuals. The striking physical resemblance of MZs, who are often confused for one another, causes their parents, their teachers and their peers to treat them very much alike.

Furthermore, "peas-in-a-pod" MZ twins tend to spend a great deal of time with each other, doing similar things—much more so than do same-sex DZ twins. These facts, established by questionnaire studies of twins, have been known for many years. MZ twins report, for example, that they have spent a night apart much less often than do DZs. MZ twins are much more likely to have the same friends, and to play together, than are DZ twins. They are also much more likely to have dressed alike. In a study by Smith published in 1965, 40 per cent of MZs reported that they usually studied together, compared to only 15 per cent of DZs. Obviously, studying the same material at the same time would tend to produce similar test scores in MZ pairs. There can be no question that, in general, MZs share more similar environments than do DZs.

More frequent meetings

A study of the possible genetic basis of dietary intake reported by Fabsitz and co-workers in 1978 underlined this point. The investigators studied a large number of middle-aged male twin pairs, all of whom had served in the United States armed forces. For such items as total calorie intake and total fat intake, the correlation among MZ twins was significantly higher than among DZ twins. That is precisely the kind of evidence which hereditarians would interpret as demonstrating the genetic basis of calorie and fat intake. The Fabsitz study, however, asked subjects the simple question, "How frequently do you and your twin get together now?" It was no surprise that, even in middle age, MZ twins reported that they saw each other much more often than did DZ twins. More interestingly, it was found that those MZ pairs who saw each other often were more alike in food intake than were those who did not. A similar difference was observed in the sample of DZ twins. This makes it reasonable to suppose that much, if not all, of the greater dietary similarity of MZs over DZs has nothing at all to do with genetics. The authors concluded: "Unequal environmental effects may lead to falsely high estimates of genetic variance for nutrient intake." Precisely the same is true, of course, of twin studies of IQ.

More similar incomes

The American veteran male twins in the Fabsitz study were drawn from precisely the same register as the male twins whose incomes had been studied by Taubman in 1976. The correlation in income of MZs was significantly higher than that of DZs. On the naive assumption that the environments of MZs and of DZs were equally similar, this finding was interpreted to mean that income was in large measure determined by the genes. When learning of this study, Professor Eysenck promptly advised a Royal Commission on the Distribution of Income and Wealth that it "might as well pack up". The results were taken to show that, since income had a genetic basis, it could not be redistributed! This is absurdly fallacious logic, even if the study had demonstrated a genetic basis for the present distribution of income, which it did not. Neither Eysenck nor Taubman had thought to ask the two types of twins how often they saw one another, or how alike their environments had been.

NON-IDENTICAL TWINS

To return to IQ studies of twins, Nichols in 1965 classified a large number of twins of high-school age in terms of whether or not they differed substantially in "similarity of experience". The DZ twins reported less similar experiences than the MZs. There was a significant tendency for female DZs to report more similar experiences than male DZs. There was no such sex difference among MZ twins. This sex difference was confirmed in 1976 by Loehlin and Nichols, who reported that female DZ twins are more likely to sleep in the same bedroom than are male DZs. The female DZs are in fact about as likely to sleep in the same bedroom as are MZ twins, among whom there is no sex difference in the tendency to sleep in the same room.

Sex differences in IQ

The fact that female DZs have clearly more similar experiences than male DZs suggests to an environmental theorist that the IQ similarity of female DZs should be greater than that of male DZs, even though DZs of both sexes have about 50 per cent of their genes in common. Typically, twin studies do not present results separately for each sex, but Kamin in 1979 listed ten studies in which female DZs resembled one another significantly more in IQ than did male DZs. There were three studies in which no such effect was observed, but in no case were the male DZs significantly more alike in IQ. There were no significant sex differences in IQ resemblance among MZ twins. These findings make perfectly good environmental—but not genetic—sense.

Perhaps the clearest example of the DZ sex difference occurs in the 1966 data of RMC Huntley. The raw data were kindly made available to me for re-analysis by Dr Huntley. The IQ correlations of Huntley's MZ and DZ twins, broken down by sex, are presented in Table 6. The correlation of female DZs is significantly higher than that of male DZs;

in this sample, it is in fact not significantly lower than that of female MZs. The one group that stands out, with a much lower IQ correlation, is the male DZs—just as, in the Loehlin and Nichols study, male DZs are the one group who stand out by tending not to sleep in the same bedroom.

Table 6. IQ correlations from Huntley (1966).

SEX:	MZ TWIN PAIRS	DZ TWIN PAIRS
Males	0.82 (N = 50)	0.51 (N = 86)
Females	0.83 (N = 45)	0.70 (N = 80)

("N" refers to the number of twin pairs on which each tabled correlation is based.)

Misclassified MZs

Possibly, however, more errors have for some reason been made among females than among males in classifying individual twin pairs as MZs or DZs. For example, if a large enough number of female pairs who were truly MZ had been mistakenly classified as DZ, that could significantly inflate the IQ correlation observed in DZs. This does not seem to have happened, however. The *same* Huntley twins, classified in the same way, had also had their heights measured. The height correlations are presented in Table 7. There is no sex difference in the height correlations, either among MZs or DZs. The correlations for height are textbook-like figures, so it is obvious that MZs and DZs have been correctly classified. The very different patterns of correlations observed for IQ and for height in the Huntley study admit of a very simple interpretation: in the case of height we are in fact dealing with a highly heritable trait.

Table 7. Height correlations from Huntley (1966).

SEX:	MZ TWIN PAIRS	DZ TWIN PAIRS
Males	0.94 (N = 50)	0.51 (N = 86)
Females	0.94 (N = 45)	0.53 (N = 80)

("N" refers to the number of twin pairs on which each tabled correlation is based.)

THE IMPORTANCE OF LOOKS

The evidence on sex differences in twin IQ correlations makes perfectly good environmental sense, but it is circumstantial in nature. What, if any, evidence do we have that those *individual* twin pairs who are treated most alike are in fact the pairs who are most alike in IQ? We do indeed know that, as a group, MZ twins are treated more alike than are DZs. Possibly, however, that has nothing to do with the higher IQ correlations of MZs. We want the answer to a simple but important question: do those MZ pairs who are treated alike resemble one another more in IQ than do MZ pairs who are not treated alike? When and if such an environmental effect can be demonstrated, it will be entirely plausible to attribute the MZ–DZ difference in IQ correlations to the greater environmental similarity of MZs.

Professor Eysenck has concerned himself with this critical question, citing primarily the data of Loehlin and Nichols. While admitting that MZ twins more often dressed alike, played together, slept in the same room, and so on, Eysenck maintains there is "not a shred of evidence" to indicate that those twins who were treated more alike were more alike in IQ. That is simply not true. The raw data of Loehlin and Nichols—on computer tape and freely available to anyone who is interested—tell a different story. Parents of twins were asked whether they had tried to treat the twins "exactly the same". Parents of MZs were much more likely to say that they had. More importantly, those MZs whose parents had tried to treat them exactly the same were found by Kamin to be significantly more alike in IQ than MZs whose parents had not. This must, of course, be an environmental effect, since all MZ pairs are genetically identical, which shows that some, at least, of the MZ–DZ difference in IQ correlation is of environmental origin. The evidence is clear in the very study Eysenck cites as demonstrating the precise opposite.

Though statistically highly significant, the difference in IQ resemblance between MZs "treated alike" and other MZs was not great. That is scarcely surprising. The reply of one parent to a single blunt question about whether twins have been treated alike is not a very sensitive measure of the similarity of twins' environments. The reply of one parent tells us nothing at all about the behaviour of the other parent, of teachers or of peers towards the twins—or about the behaviour of the twins towards each other. Presumably, more accurate and sensitive measurements of the twins' environments could reveal much larger effects.

Blunt measurements

The bluntness of environmental measuring in twin studies may be largely responsible for investigators' failure to detect environmental effects. For example, Professor Scarr, in her 1980 Philadelphia twin study, asked her subjects whether or not they usually dressed alike. Responses were scored on a blunt two-point scale: yes or no. There was no relation, the study reported, between a twin pair's similarity in IQ and whether or not they dressed alike. The twins, however, had actually responded to the question about dress on a four-point scale. The alternative answers had been "almost always", "frequently", "some-times", and "seldom". From Professor Scarr's raw data it can be calculated that 17 MZ pairs agreed that they almost always dressed alike, while 43 pairs agreed that they seldom did so. MZ twins who always dressed alike were significantly more similar on the non-verbal Raven Progressive Matrices test. Translating from standard deviations to the IQ equivalents, MZs who dressed alike differed by an average of 5.7 IQ points. Those who seldom dressed alike differed by about 10.7 points. MZs who did not dress alike were not much more similar in IQ than are ordinary siblings. There is nothing very sensitive about a four-point self-report scale on the subject of dressing alike; but even so blunt a measure of environmental similarity can reveal effects operating to produce IQ resemblance in MZ twins.

The greater the resemblance, the closer the IQs

The major factor that results in MZ twins experiencing such similar environments is their striking resemblance in physical appearance. While most MZ twin pairs look very much alike, some pairs look less alike than others. An environmentalist would obviously expect those MZ pairs who look most alike to be most alike in IQ as well. The raw data from the Philadelphia study by Professor Scarr can be used to test this prediction.

The twin pairs in the Philadelphia study were rated for similarity of physical appearance by a group of eight judges, who made their ratings on a six-point scale by comparing twins' photographs. For 121 MZ pairs, there was a significant correlation of 0.26 between difference in appearance and difference in Raven's Progressive Matrices test. That is, MZ twin pairs who looked most alike were also most alike in IQ. Taken as a whole, MZs are of course much more similar in appearance than DZs; so again, we must conclude that at least part of the MZ–DZ difference in IQ correlations is environmentally produced. The fact that MZs look so much alike is of course genetically determined—and that they are so alike in IQ appears to be an indirect and non-genetic consequence of this. To regard the difference between MZs and DZs in IQ correlation as an index of the heritability of IQ is an error.

DZs AND SIBLINGS

We have stressed that MZ twins have more similar environments than DZs, and that this can account for the higher IQ correlation of MZs. We must recognise, however, that even DZ twins experience more similar environments than do ordinary siblings, who, unlike DZ twins, are born at different times, have different age mates, and so on. An environmentalist would expect a higher IQ correlation among DZ twins than among ordinary siblings—even though the degree of genetic resemblance is the same in each case. Few studies have examined both DZ twins and ordinary siblings, but the available evidence clearly supports the environmentalist view.

The fraudulent "adjusted assessments" of Cyril Burt, as we might expect, indicated no difference at all between correlations for DZ twins and for siblings. The genuine data of Herrman and Hogben produced correlations of 0.49 for DZs and 0.32 for siblings. And Tabah and Sutter reported 0.58 for DZs and 0.45 for siblings. Part of this apparent difference between DZs and siblings may be caused, however, by imperfect age standardisation of IQ tests. DZ twins, unlike ordinary siblings, are usually tested at the same age; if the test is not perfectly standardised, this will tend to inflate the correlation for DZs relative to that for siblings.

Well controlled and relevant data do exist, however, provided by a pair of studies by Record, McKeown and Edwards. The children in their sample—both twins and siblings—were *all* tested as they reached the age of 11. The sample included 358 pairs of opposite-sex (and thus obviously DZ) twins, for whom the IQ correlation was 0.62. This is significantly higher than the 0.55 correlation observed among 2,525 pairs of opposite-sex siblings. This study controls for age effects—that is, the possibility of age affecting results is taken into account in the design of the study. The restriction to opposite-sex twins and siblings not only controls for possible sex differences; it ensures that all twins classified as DZs are in fact true DZs. The results appear conclusive, then. The greater similarity of environment experienced by DZ twins does make them more alike in IQ than ordinary siblings—even though DZ twins and siblings are, genetically, equivalently similar. We might imagine that same-sex DZ twins would experience even more similar environments, and thus be even more alike in IQ; but Record, McKeown and Edwards did not classify their same-sex twins into MZ and DZ groups, so we cannot be certain.

SAME-SEX AND OPPOSITE-SEX TWINS

What data do we have comparing same-sex and opposite-sex DZ twins? From an environmentalist view, it seems reasonable to suppose that the greater shared experience of same-sex twins would result in a higher IQ correlation. There are, again, not many studies that make the relevant comparison; but at least four studies of school-age children do so. The results of those four studies are summarised in Table 8. Three of them indicate a significantly greater resemblance in IQ among same-sex DZs.

Gross bias in selection

The only exception is the report by Herrman and Hogben, which suffers from a peculiar deficiency: they had a great deal of difficulty in determining which pairs were MZ and which DZ. The study was conducted before modern blood testing techniques greatly improved the reliability of such diagnoses. A full 28 per cent of the same-sex twins studied by Herrman and Hogben could not be diagnosed, and the same-sex DZ twins finally included in the study were especially selected for their "striking physical dissimilarities". There was no such selection bias in the case of opposite-sex twins, who could all, of course, be recognised at once as DZ. The excluded same-sex pairs who looked alike were presumably the very pairs who had experienced the most similar environments, and would have had the most similar IQs.

The gross sampling bias in the Herrman and Hogben study is sufficient to explain its different outcome. The weight of evidence from studies comparing same-sex and opposite-sex DZs clearly supports the environmentalist expectation. The difference observed might partly result from incorrectly including some true MZs in the sample of DZs; but it does not seem likely that the entire effect could be explained in this way.

Table 8. IQ correlations comparing same-sex and opposite-sex DZ twins

STUDY:	SAME-SEX DZs	OPPOSITE-SEX DZs
Stocks and Karn (1933)	0.87 (N = 27)	0.38 (N = 28)
Herrman and Hogben (1933)	0.47 (N = 96)	0.51 (N = 138)
Huntley (1966)	0.66 (N = 135)	0.45 (N = 100)
Adams et al. (1976)	0.66 (N = 55)	0.47 (N = 40)

("N" refers to the number of twin pairs on which each tabled correlation is based. The correlations tabled for Adams et al. are the means for two separate tests, verbal and non-verbal.)

Professor Eysenck has recently informed his readers—incorrectly—that "there is no difference between like-sexed and unlike-sexed DZ twins" and that "DZ twins are no more alike than ordinary full siblings". Eysenck used these claims in an attempt to refute the environmentalist critique of IQ studies of MZ and DZ twins. The only study quoted by Eysenck in support of his two incorrect claims was the "early, carefully planned" work of Herrman and Hogben.

This review of MZ–DZ twin studies, like our earlier reviews of separated MZs and adoption studies, has failed to reveal any unambiguous evidence for the heritability of IQ. We have observed once again that evidence apparently consistent with a genetic interpretation is equally consistent with an environmental interpretation. From either viewpoint, one expects the correlation of MZs to be higher than that of DZs. The environmentalist view, however, correctly predicts that those MZs who are most alike in appearance, and who have been treated most similarly, should be most alike in IQ. The environmentalist view also correctly predicts sex differences observed among DZ twins, and the difference between DZs and ordinary siblings. These findings cannot disprove the possibility that some part of the MZ–DZ difference in IQ correlation is a genetic effect—but they do show, at the very least, that any estimates of heritability derived from twin studies are inflated.

17
KINSHIP
CORRELATIONS
AND THE
MODELLING GAME

*"It would be true to say that their work has revolutionised the field. . . .
The careful analyses performed by Jinks and Fulker and their colleagues
at Birmingham are models of what genetic work should be. . . . It is
these quite recent methods and models . . . which justify us in asserting
that the heritability of intelligence is approximately 0.80. Critics of this
figure would have to tangle with the geneticists who have wrought this
revolution in analysis, not with psychologists who may claim to
understand, but not to have originated, this important advance."*

HANS J EYSENCK, 1973

There is no escape. With trepidation we must now tangle, if not with
geneticists, at least with the models which they construct and psychologists claim to understand. We had better begin by pointing out two
simple facts. The first is that models, no matter how ingenious they are,
must be applied to actual data collected in the real world. The
mathematical cleverness of a model can in no way compensate for data
that are falsified, biased or of poor quality. The second is that even a
close fit between a particular model and genuine data cannot prove that
the model is "true". The particular model might, for instance, emphasise
genetic factors, and it might fit the data very well. That would not mean
that a different model emphasising environmental factors could not fit
the data just as well, or better.

There are varying degrees of closeness of kinship within human
families, and the closer the degree of kinship between two individuals
the more genes they will share in common. Thus, under a genetic model,
one would expect close relatives such as parent and child to have a higher
IQ correlation than, say, grandparent and grandchild. The same
prediction, however, could also be made under an environmental model,
because the more closely related two individuals are, the more similar

their environments are likely to have been. To test models, in any event, it is necessary to have a number of *kinship correlations* for IQ—that is, correlations obtained from relatives of varying degrees of closeness. The effect of family environment can in theory be assessed by including in the set of correlations to which the model is fitted data for relatives, such as identical twins and adopted children, who have been reared apart.

An above average resemblance between members of the same family. Twin studies show significant correlations between difference in appearance and difference in IQ but the environmental view is that similar mental abilities develop because of similar, physiologically determined, experiences.

PATCHWORK CORRELATIONS

We have already noted that only one investigator, Cyril Burt, ever claimed to have collected a full set of kinship correlations, using the same test in a single population. His "adjusted assessments", must, of course, be discarded for the reasons we have specified. The alternative adopted by most model-fitters is a piecemeal approach: they use as their data either the *average correlations* or the *median correlations* for each kinship category given in the reports of several different investigators. (With a series of numbers, the median is the middle one, whereas the average, also called the mean, is the total divided by the number of items in question. In the series 3, 6, 12, for instance, the median is 6, while the average is 7.) Studies of this sort have employed very different IQ tests, administered to different populations at different times on different continents. The model fitted by one geneticist may omit, or include, in its averaged correlations studies included or omitted by other model-fitters. This, of course, can make a considerable difference.

The most influential collection of median kinship correlations was provided in 1963 by Erlenmeyer–Kimling and Jarvik. They combed through 52 separate studies, performed in eight different countries, to gather IQ correlations for a total of ten different kinship categories. The chart summarising their work has been reproduced in countless textbooks and monographs. Their report has been described by Vandenberg, a leading behaviour geneticist, as "a paper that condensed in a few pages and one figure probably more information than any other publication in the history of psychology". The apparent correspondence between the median correlations reported by Erlenmeyer–Kimling and Jarvik and those derived from a "genetic model" has been stressed by many hereditarians, including Burt, Jensen and Eysenck.

The median values calculated by Erlenmeyer–Kimling and Jarvik are derived, however, from a set of studies with chaotically varying individual values. Erlenmeyer–Kimling and Jarvik were able to locate 12 different studies conducted before 1963 that reported a parent–child IQ correlation. The median of the 12 reported values was precisely 0.50—a number beautifully consistent with a very simple genetic expectation. But the values reported in individual studies varied all the way from about 0.20 to about 0.80. Kamin, reporting in 1979, located 16 studies conducted *since* 1963, in which the median parent–child correlation was only 0.33, with individual values ranging from 0.08 to 0.41. Peculiarly, not a single study after 1963 has reported a value as high as the median reported by the pre-1963 studies. The median parent–child value of 0.33, if inserted into the set of correlations to which genetic models have been fitted, would wreak havoc on the models. They would no longer fit. The truth is that we simply do not know the "true" value for the parent–child correlation, or for any other, which in itself is sufficient to invalidate the attempts to apply models to median or average kinship correlations.

The Burt and Howard model

The most influential early model to which kinship correlations were fitted was originally developed by Ronald Fisher in 1918. The Burt and Howard model, applied to the median kinship correlations of Erlen-meyer–Kimling and Jarvik, indicated an IQ heritability of over 80 per cent. The close fit of the model to the median correlations was pointed to with pride by Jensen, Eysenck and other hereditarians.

To apply the Burt and Howard model, at least three kinship correlations are needed—those for husband and wife, for parent and child, and for siblings. Three recent family studies—those of DeFries, 1979, Spuhler, 1976 and Guttman, 1974—qualify for this treatment. These studies did not, it must be noted, use median correlations from different studies, but the results of their own testing. In all three studies, as it happens, subjects were given the same IQ test—Raven's Progressive Matrices, a non-verbal test often described as an almost pure measure of a hypothetical "general intelligence factor". The model fails with each of the three studies. That is, it produces mathematically impossible degrees of absence of genetic dominance. Were we to overlook this fatal defect, the model would indicate broad heritabilities of 25, 13 and 26 per cent in the three studies respectively. The Burt and Howard model, in short, fits (and suggests very high heritability) only when it is applied to the artificial median correlations, or to Burt's "adjusted assessments", not when it is applied to real-world data collected in individual controlled studies.

THE NEW GENERATION OF MODELS

The Burt and Howard model is anyway old-fashioned now. Much more sophisticated and mathematically complex models are available. The basic feature of more recent models is that they employ appropriate statistical techniques to fit, simultaneously, the whole range of kinship correlations—not just the three kinships used by Burt and Howard—producing the best possible overall fit, balancing off discrepancies as well as possible, and giving due weight to the sample sizes of different kinships. How well the model and the data fit can be tested statistically, and estimates of the heritability of IQ can be arrived at. There are two major schools of model-makers, each with its own assumptions and techniques—one in Birmingham, England, the other in Honolulu.

An obvious problem faces today's ambitious model-makers. Their models are meant to be applied to a full range of kinship correlations—but alas, nobody has collected a full set of such correlations in a single study. Thus the models have been applied, by different model-makers, to various arbitrary sets of median or average correlations. Though the models are new, the average correlations to which they have been applied are not. The post-1963 parent–child median of 0.33, for example, has not been used in any of the model-fitting exercises. Nor have the recent family studies, whose results are inconsistent with high heritability, been used. If they were to be thrown into the pot when computing average correlations, they would be swamped by a motley collection of statistics gathered between 1912 and 1963.

To understand in detail the results of the model-fitting exercises, some knowledge of quantitative methods is essential. Fortunately, the American econometrician Arthur Goldberger has summarised work in this field in an especially lucid and informative way. He has not hesitated "to tangle with the geneticists who have wrought this revolution in analysis", and we can now follow his critical comments.

The Birmingham school, in 1970 work by Jinks and Fulker and 1975 work by Eaves, applied models to two different sets of English kinship correlations. The models fitted the data closely and suggested heritabilities of 83 and of 85 per cent. The problem is that all the correlations used in these two exercises were the "adjusted assessments" of Cyril Burt. The fact that a genetic model closely fits Burt's fraudulent data is no cause for celebration. The Birmingham type of model was then applied to two different sets of American correlations—by Eaves in 1975, and again in 1977. The calculated heritabilities now became 68 and 60 per cent. The agreement between these two values is more apparent than real, however. The Birmingham model—unlike the Honolulu model—decides in advance to allow an effect of genetic dominance to emerge from the analysed correlations. Using the 1975 figures, about half of the genetic variance was the effect of dominance, while with the 1977 figures literally none of it was. With the 1975 figures, virtually all the environmental variance in IQ was said to be shared by members of a family; with the 1977 ones, literally all environmental variance was unique to individuals, none being shared by family members.

Jencks applied a complex and somewhat jerry-built model to a set of American median correlations and reported an IQ heritability of only 45 per cent in 1972. The Jencks model—unlike that of the Birmingham school—allowed for the possibility of gene–environment *covariance*. That is, people inheriting "high-IQ genes" might be just those people who also tend to experience environments which favour the development of high IQ—which, if it is the case, would make it impossible to assign their high IQ either to genes or to environment. The arbitrary model used by Jencks concluded that about 20 per cent of the total variance of IQ was due to just such gene-environment covariance.

The Honolulu type of model was applied to a set of American median kinship correlations by Rao, Morton and Yee. This model, like that of Jencks, allows for an effect of gene-environment covariance, which was found in this case to account for about 14 per cent of total IQ variance. But unlike the Birmingham and Jencks models, the Honolulu model does not allow for an effect of genetic dominance, so of course none was found.

Their model has the interesting feature of making it possible to compute heritabilities separately for children and for adults. Rao, Morton and Yee estimated a heritability of 67 per cent among children, and of only 21 per cent among adults, and suggested that, as people leave school and enter specialised occupations, variation in IQ might become more and more dependent on differing experiences. The same type of model was applied to an expanded set of American correlations, with basically similar results.

Built-in false assumptions

There is, as we have seen, much that is arbitrary in the construction of a model—and there is no one to say that the assumptions of one model-maker are more scientifically valid than another's. There are some clearly invalid assumptions, however, built into all models—assumptions that have the effect of inflating estimates of heritability. The Birmingham and Honolulu models share the wholly incredible assumption that, in terms of relevance to IQ, the experiences of a pair of MZ twins are no more similar than those of a pair of ordinary siblings. Both schools also make the demonstrably false assumption that adopted children are placed into families randomly drawn from the general population and exhibiting the full range of environmental variation. The absurd nature of these assumptions has been documented in the two previous chapters.

These false assumptions are not merely incidental. They are absolutely central to the game played by model-makers. The fact that the assumptions imbedded in the models are unbelievable means that the conclusions cranked out of the models cannot be believed either. The reason for this unhappy state of affairs has been spelled out plainly by Professor Goldberger:

"To explain the persistent use of such assumptions, we need only recognize that without them the models would be indeterminate. If less restrictive, and hence more plausible, specification were made, the number of unknown parameters would approach and soon exceed the number of observations. Implausible assumptions are needed to identify the parameters and produce the estimates, and thus to keep the model-fitters happy. But estimates produced in that manner do not merit the attention of the rest of us."

Professor Eysenck has told his readers, glowingly, of a "revolution in analysis" brought about by geneticists—one that psychologists might understand, but presumably beyond the grasp of laymen. We have been forced to note, however, that there are no clear and reliable data to which these revolutionary models can be applied. We do not even know what the true correlation between parent and child is; studies reported since 1963 suggest a value that could not be fitted to the models. Perhaps more important, we have noted that competing models are based on very different assumptions—and that all the models share some convenient, but demonstrably false, assumptions. To appreciate the pleasure that model-fitters derive from playing the modelling game, it is necessary to be skilled in quantitative matters. The lay person, however, can easily understand that such aesthetically pleasing models do not tell us anything about the real world. The important thing is not to be blinded by appeals to authority or by complex-looking formulae—or by loose talk about scientific revolutions.

18
SOME
ODDS AND ENDS

"As an example, take the Irish—a well defined, interbreeding population, isolated on an island . . . subject to historical processes which . . . have drawn away . . . the most able and adventurous of citizens to foreign countries. Under these circumstances . . . we might expect a distinctly lower IQ level among the remaining Irish. . . . Facts seem to confirm these hypotheses; . . . the Irish . . . have IQs . . . not very different from those observed in American negroes, and far below comparable English samples."

HANS J EYSENCK, 1971

We shall discuss in this section three separate matters, none of them central to the question of IQ heritability. The first topic, the average IQ difference between American blacks and whites, has nevertheless inspired most of the present interest in heritability and IQ. The second topic, the phenomenon known as *inbreeding depression*, would, if demonstrated, provide unique evidence for the genetic basis of IQ. There are some theorists who maintain that the lowering effect of inbreeding on IQ has indeed been demonstrated, so we will review, critically if briefly, the available evidence. Finally, several hereditarians have asserted that whatever small effect environment may exert is largely prenatal—not social, cultural or educational. The pathetically inadequate evidence for intra-uterine influences on IQ will be discussed in the final pages of this chapter.

BLACK-WHITE DIFFERENCES IN IQ

The clear fact that, on average, American blacks have lower IQ scores than American whites has been well known since the American army's IQ testing programme in the First World War. The difference is about 15 IQ points. The fact is not in dispute: the argument has revolved around how to interpret that difference. To most psychologists and social scientists, the obvious educational, social and economic discriminations to which blacks have been subjected seem entirely adequate to explain the difference in measured IQ. There is no doubt, however, that black and white groups living in America do differ in genetic make-up so that there is a theoretical possibility that some part of the observed difference could be genetic. There is also a theoretical possibility, it should be noted, that the genes of blacks could produce *higher* average IQ scores than those of whites if both races lived in equally favourable environments.

The research on race differences has inevitably been inconclusive; and the uses to which such research has been put have often been distinctly unsavoury. For those with the stomach for it, the relevant research was reviewed in detail by Loehlin, Lindzey and Spuhler in 1975. We shall discuss the research here only in broadest outline.

A whiter shade of black

The First World War data indicated that the average IQ of blacks in some northern states was higher than the average IQ of whites in some southern states, findings which seemed consistent with the higher educational and economic standards of the North. The hereditarians countered by arguing, without evidence, that genetically superior blacks had selectively migrated from southern to northern states. The hereditarians demonstrated that darker-skinned blacks had lower average IQs than lighter-skinned blacks. This, it was argued, was due to the greater proportion of Caucasian genes supposedly inherited by light-skinned blacks. The environmentalists countered by pointing to the obvious fact that light-skinned blacks suffered less discrimination. The hereditarians suggested that studies be undertaken of the relation between IQ and the proportion of Caucasian genes, estimated by blood groups. When these studies found no relation between IQ and the proportion of Caucasian genes inherited by individual blacks, the hereditarians who had suggested them concluded that blood groups

Black US serviceman: discriminated against at home but in the front line of the war in Vietnam.

provided a poor estimate of the proportion of Caucasian ancestry. When it was observed that children born out of wedlock to white mothers and black fathers had higher IQs than children born to black mothers and white fathers, it was suggested that the black fathers involved in inter-racial mating might have been more intelligent than the black mothers. When it was demonstrated that black children adopted into upper-SES white homes developed distinctly superior IQs, it was said that this was not inconsistent with *part* of the normally observed black-white difference being genetic.

There is obviously no point in continuing along these lines. There is no way of providing a definite answer to the question of black-white differences until and unless we are able to build a society in which blacks and whites are exposed to similarly favourable, and non-discriminatory, environments. The irony is that if we succeed in building such a society, nobody will any longer be interested in answering the question. The causes of differences in racial averages seem important only in a society obsessed with racism. Group differences in average IQ tell us nothing, of course, about individuals. And even if differences in group averages could be shown to be partly based on genetic differences, it would not mean that environmental intervention could not modify—and even reverse—the difference. For those obtuse enough to have doubted the importance of environment, the transracial adoption study of Scarr and Weinberg demonstrated that black children adopted into advantaged white families at under one year of age developed an average IQ of 110. We do not—for obvious reasons—have data on the IQs of white children adopted into disadvantaged black families. There seems little doubt about what the outcome of such a hypothetical study would be.

IQ and information processing

The attempt to demonstrate that blacks are genetically inferior moved on to new ground in a study briefly reported by Jensen in 1975. It involved a group of blacks and a group of whites who, Jensen asserted, had been matched for having the same average IQ. To assess mental functioning, the subjects were all required to press a button as rapidly as they could when a signal light flashed on. When the task involved responding to only one light, there was no difference in the speed with which blacks and whites were able to press a button next to the light. When the task was to respond by pressing a button next to one of several possible lights, whites were said to respond more rapidly than blacks. The superiority of the whites increased as the number of possible alternative lights increased. These remarkable data supposedly demon-strated that whites could "process information" more efficiently than blacks, even if the individuals involved had been matched on standard IQ measures. The Jensen data are reminiscent of Cyril Burt's 1911 report that, while Liverpool slum boys and Oxford preparatory (private) school boys performed equally on *simple* sensory tasks, the Oxford boys excelled on *complex* sensory tasks. These results, Burt believed, "harmonised with the results of comparative investigations upon savage and civilised races".

The Jensen data on reaction times seem to me inherently implausible. What conclusions can we draw from his supposed demonstration that blacks are inferior to whites in the speed with which they respond to complex visual information? We would have to conclude that in occupations which demand fast response to complex visual displays—for example, professional boxing and basketball—blacks would be underrepresented. This is obviously not the case, and it seems to me that any sensible person would conclude that an error of some kind has found its way into Jensen's data or procedures. In the same 1974 article in which he admitted that Burt's data were useless, Professor Jensen wrote of the obligation on scientists to make their raw data available for re-analysis to other interested scientists. Thus I assumed that when I asked him for the raw data of his reaction time study, offering to defray all expenses, he would routinely oblige. Though I have repeated my request on several occasions, indicating that my purpose was to search for possible errors, Jensen has continued to refuse to provide the data. His refusal does little to inspire confidence in the objectivity of his race-oriented research.

Past kings of the boxing world, Mohammed Ali and George Forman. Tiptop reflexes.

INBREEDING DEPRESSION

Professor Eysenck (1973) has asserted that "lowering of IQ after inbreeding is perhaps among the best lines of evidence we have for hereditary control of intelligence." Fuller and Thompson came to a different conclusion in their influential text on behaviour genetics: "In general, then, the data on consanguinity or inbreeding effects on IQ are rather meager and ambiguous. Few firm conclusions can be educed from them." This is not the only occasion on which Professor Eysenck's judgment has differed from that of more sober-minded authorities. The data on inbreeding depression have been reviewed in detail elsewhere (Kamin, 1980). For present purposes, only a few brief remarks seem necessary.

The phenomenon of inbreeding depression occurs when a trait is determined by the action of many genes and when, as is often the case, genetic dominance favours a high value of the trait. This is said by hereditarians to be the case with intelligence. The effects of unfavourable recessive genes are normally suppressed by the action of dominant genes with which they are paired. The offspring of genetically related mates, however, will tend to receive the same gene twice from the same ancestor, and harmful recessive genes will be more likely to occur in pairs. Their unfavourable effect will not be suppressed by a dominant gene, and the trait will take on a lower value. This is inbreeding depression. The demonstration of a lowering of IQ as a result of inbreeding would be unique evidence that IQ is in fact under some hereditary control.

Cousin marriage in Japan

The best known and largest study of IQ and inbreeding depression was conducted in Japan by Schull and Neel, who reported in 1965. The practice of cousin marriage is relatively common in Japan, and Schull and Neel were able to study 865 children of such marriages. The inbred children were compared to a control group of children of unrelated parents, sampled from the same population. Following a complicated series of statistical adjustments, Schull and Neel estimated that children of first cousins should show a "modest" lowering of IQ of about 4.5 points compared to control children.

The complicated statistical adjustments and estimates were made necessary by an unanticipated finding. When Schull and Neel measured the SES (socio-economic status) of their subjects, they found that cousins who had married were of significantly lower social class than the unrelated couples. The children's IQ scores were related to their SES, quite apart from any inbreeding, with lower-class children having lower IQs. The children of cousin marriages would therefore be expected to have lower IQs merely because of their lower social class. Though Schull and Neel *estimated* that there would still be some lowering of IQs as a result of inbreeding even if SES had been taken into account, that is merely conjecture. With a more accurate measurement of SES, they admitted, the entire effect attributed to inbreeding depression might well vanish.

Several other studies have failed to find significant lowering of IQ from inbreeding, but Bashi, reporting in 1977, claimed to have observed a "modest" but significant effect among Arab children in the Israeli school system—about one or two IQ points for the children of first cousin marriages. The study, however, was much less painstaking in its analysis of SES effects than the work of Schull and Neel. It was based on a national sample. There is some likelihood that cousin marriages are more common in country villages than in cities, and that IQ scores are lower in villages than in cities, which, if true, would mean that children of cousin marriages would have lower IQs for reasons having nothing to do with inbreeding depression. Without much more extensive data on the general population than are available, there are no adequate grounds to attribute the modest effect reported by Bashi to inbreeding depression.

THE UTERUS AND IQ

Hereditarian theorists have persistently argued that by far the greatest part of IQ variation—about 80 per cent—is genetically determined. That, even if true, would leave 20 per cent of IQ variation to be determined by environment. Some hereditarians point out that "environment" does not necessarily mean anything social, cultural or educational; that the first environment encountered by the developing foetus is, after all, its mother's uterus, and that the 20 per cent of IQ variance determined by environment might be largely the result of intra-uterine conditions, including inadequate prenatal nutrition, birth trauma and so on. This kind of suggestion has been made explicitly by Jensen, and by many others. To critics with a suspicious cast of mind, it seems that hereditarian theorists will go to almost any lengths to deny the importance of social and educational factors.

Broman, Nichols and Kennedy conducted a large-scale study into the influence on IQ of conditions in the womb and during birth and could find no significant effect. What little evidence there is of an effect comes from studies of identical twins. Though genetically identical, MZ twins have different birth weights. We might assume that the heavier twin has had a better intra-uterine experience, and there is indeed evidence in some (though not all) studies that the heavier of a pair of MZ twins at birth tends to develop the higher IQ. But the difference, if any, is small. For the remarkable fact is that, in studies which report such an effect, the *size* of the difference in birth weight is of no consequence at all: a twin only a couple of grams heavier than its mate tends to have the same IQ advantage as a twin many hundreds of grams heavier than its mate. This ungraded effect makes little biological (or other) sense. To say the least, it is difficult to interpret.

The evidence of transfusion syndrome

The subject of MZ twins' birth weights was brought into prominence in 1977 in an article by Professor Harry Munsinger about a condition known as transfusion syndrome. Transfusion syndrome, which sometimes affects MZs, involves blood transfer between the twins in the uterus. One twin is in effect an involuntary blood donor to the other and may as a result weigh less at birth. The poor intra-uterine experience of the donor twin might be thought to injure its brain and thus depress its IQ. Since this chain of events, which would inflate the IQ difference between a pair of MZ twins, is environmental, Munsinger argued that the IQ correlation between MZs might underestimate the powerful role of the genes; if it were not for transfusion syndrome and intra-uterine experiences, MZ twins might have identical IQs. To test this possibility, Munsinger re-analysed the data on MZ twins, attempting to separate pairs in which transfusion syndrome had occurred from pairs in which it had not. A large difference in birth weight was assumed to indicate transfusion syndrome. Munsinger's conclusions were truly extraordinary: "Postnatal social and cultural environmental influences have no effect on the population variation in IQ." The heritability of IQ was said to be about 95 per cent. With more reliable IQ tests, the heritability would be 100 per cent.

Munsinger's comedy of errors

These incredible conclusions were taken seriously by authorities one might have thought should know better. The American Nobel Prize-winner William Shockley told *The Times Higher Education Supplement* that he doubted whether Burt's results were faked and was impressed by the work of Munsinger on transfusion syndrome. Professor Eysenck wrote to a scientific magazine to explain that, even if Burt's data were discarded, the evidence for high heritability of IQ would be overwhelming. The Munsinger analysis of transfusion syndrome was cited as an important source of such evidence. Munsinger's analysis, however, can only be described as a comedy of errors, as I have pointed out in a critical article published in 1978 in the journal which carried Munsinger's original paper. Here is a list of just some of Munsinger's errors.

To begin with, Munsinger in many cases *guessed* at the birth weight of twins from vague verbal descriptions in published case studies. His guesses systematically biased the results in favour of his hypothesis. Stated birth weights were simply ignored in some cases which worked against his hypothesis. There were also guesses (and errors) made in tabulating some IQs. Munsinger made large errors in his own favour when he transcribed for the purpose of re-analysis the birth weight and IQ figures clearly set out in studies by other researchers. Finally, working with incorrect numbers, Munsinger applied an incorrect statistical formula. When the actual numbers from the studies re-analysed by Munsinger were subjected to the proper statistical formula, there was no support at all for his claims. The IQ correlation of MZ twins was the same whether or not cases of transfusion syndrome were included in the analysis.

There have been no recent references to Munsinger's work by those authorities who, when Burt was exposed, promptly cited Munsinger as an important source of evidence. The absurd claims of Munsinger's paper, based upon transparent and easily documented errors, were swallowed uncritically by the "experts". Like their original acceptance of Cyril Burt's absurdities, this reflects unfavourably on their critical abilities, or at any rate on the way in which they exercise them.

19

FACTS, WISHES AND EYSENCK'S REFERENCES

"Let [readers] be reassured that I have consulted not only these books, but above all the primary references therein given . . . On a few occasions I have mentioned individual references . . . but for the most part the reader will have to rely on the general watchfulness of my colleagues to make sure that I have not tried to slip anything over on him."

HANS J EYSENCK, 1971

Professor Eysenck calls attention with these words to a very serious problem. The general reader—like most scientific readers—has not the time, the resources, or the inclination to read the thousands of research reports concerned with IQ and heredity. Most of us must depend for our knowledge about such matters on books and reviews churned out by "experts" and "authorities" who invariably maintain that they are objective and concerned only with the search for truth but often have deeply-held opinions. What guarantee do we have that the "facts" they assert to be true have not been distorted to fit their views and prejudices? The wish, after all, is often father to the thought, and to the belief. Why not to the "fact" as well?

READER BEWARE

The most influential and widely read reviews of the research literature on IQ have been written by ardent hereditarians. The facts cited in those reviews tend to be as fiercely selective as those cited by a lawyer arguing his case in court. This is perhaps inevitable in such an ambiguous and emotional area. What is less excusable is the tendency for hereditarian writers to be not just selective but grossly inaccurate. Professor Jensen and his American supporters have repeatedly misrepresented simple and documentable matters of fact—and the misrepresentations always have the effect of building up, falsely, the hereditarian case. This assertion has been documented in detail in my 1974 book. The present chapter takes up Professor Eysenck's invitation to be watchful and examines the accuracy with which he has represented facts to his readers in the past.

We shall see that Eysenck has managed to slip a good many matters over, not only on his readers but also on himself.

Slipshod references

Professor Eysenck goes out of his way to assure his readers that he has consulted relevant books, and above all the primary references they contain. This scholar doth protest, perhaps, too much. There are telltale signs that Eysenck has cut an occasional corner while reading his primary references. To cite a trivial example, in 1971 he referred to Barbara Burks, author of an influential adoption study, as "he". This failure to notice the author's name had been corrected by 1973, when Eysenck properly referred to Burks as "she". But the corrected reference to Burks was supplemented by a reference on the same page to the adoption study by Alice Leahy—unfortunately, described as "he".

Failure to think very deeply about what he has read can be detected in many of Eysenck's references. The objection that adoption studies are contaminated by selective placement is dismissed with the following riposte. "That the criticism is without value can be shown by looking at the actual homes of the separated identical twins studied by Sir Cyril Burt . . . .The correlation between the socio-economic status of the home in which the one twin was brought up correlated 0.03 with that in which the other twin was brought up" (1971). The bad grammar, perhaps the result of hasty writing, can be forgiven; but Professor Eysenck had not—any more than Burt did—looked at the "actual homes" of separated twins. The incredible lack of selective placement that Burt claimed to have observed was naively accepted by Eysenck—who failed to notice that the numbers Burt used to support his assertion were flatly contradicted by the numbers in an earlier Burt report.

Misleading claims

Eysenck's tendency to exaggerate is nicely illustrated by his reference to the 1931 orphanage study by Lawrence. His argument is that, since children reared in an orphanage are all exposed to pretty much the same environment, they should—if environment is very important—all develop about the same IQ. The standard deviation in their IQ should be relatively small. There are serious defects in Lawrence's study, but for argument's sake let us grant Eysenck's assertion that any reduction in the IQ variation of the orphanage children, attributable to the similarity of their environment, was very small. From this starting point, Eysenck arrives at profound social and political conclusions: "The minute shrinkage in variance found in this study could not be increased in any political regime, however egalitarian, because it is difficult to see how such a regime could provide an environment less varied than that found in an orphanage" (1979).

A first-hand look at the Lawrence report shows that the environment to which the children were exposed was far from standardised. To begin with, the children were not given up by their mothers until they were six months old on average. Then up to the age of five or six years, they were boarded out in individual foster homes. The foster mothers, Lawrence says, continued to keep in touch with the children after they were placed in the orphanage. The children's environments, then, varied considerably for at least the first six years of life, by which time IQ—as both hereditarians and environmentalists recognise—has typically become fairly stable. There is therefore no particular reason to expect a reduction in IQ variance among the children studied by Lawrence. The point, of course, is not to advocate the imposition of some standardised environment by any regime. The point is to show that Eysenck's claim that it is difficult to see how a less varied environment could be devised is absurdly misleading. The reader who reads Eysenck first, then Lawrence might indeed conclude that something has been slipped over on him.

A fictitious matching

There are flights of fancy, masquerading as references to research literature, which are more than mere exaggerations. They are simply false. Thus to demonstrate that inbreeding depresses IQ, Eysenck cites the 1965 work of Schull and Neel as follows: "The parents [of inbred children] were carefully matched with a control group of equal age and socio-economic status" (1973). That is not the case. The inbred and control children in the Schull and Neel study, as the authors stated, differed significantly in terms both of parental age and of parental SES—and in other ways, as well. This failure to match the two groups for SES, as we pointed out in the section on inbreeding in Chapter 18, is absolutely critical. The low IQ scores of inbred children can be attributed to their low SES rather than to inbreeding. This embarrassment to his theory is neatly sidestepped by Eysenck, who invents a matching that did not take place, assuring his readers that it was performed "carefully".

Where is the evidence?

On other occasions, without citing any references at all, Eysenck makes convenient claims that cannot be supported by actual data. In his efforts to demonstrate that IQ is something more fundamental and biological than mere school performance, Eysenck makes the following undocumented assertion: "Tests of school achievement show evidence of much lower heritability than do measures of IQ . . . The degree to which achievement depends on heredity, as suggested by tests of this kind, is at most half that demonstrated for IQ, and usually it is much less" (1971). Few, if any, knowledgeable workers in the area would accept this. Perhaps Eysenck has once again been misled by the claims of Cyril Burt, who greatly emphasised the distinction between IQ tests and tests of "scholastic attainment". Burt presented data for a full range of kinship categories to show that having been reared together is more likely to make two children similar in school attainment than in IQ. The tests of IQ and of school attainment were unspecified; and the correlations remained identical to the third decimal place as sample sizes fluctuated up or down. Burt's data appear, nevertheless, to have convinced Eysenck.

Professor Eysenck informs his readers—again without references— that "African negro children (and American negro children as well) show highly precocious sensorimotor development, as compared with white norms" (1971). The reader is immediately warned, however, not to draw improper inferences. "The observed precocity lasts for about three years, after which time white children overtake the black ones. These findings are important because of a very general law in biology according to which the more prolonged the infancy, the greater in general are the cognitive or intellectual abilities of the species. This law appears to work even within a given species; thus sensorimotor precocity in humans, as shown in so-called 'baby tests' of intelligence, is negatively correlated with terminal IQ."

This is a truly stunning *tour de force*. First it is said that blacks show precocious physical development. That does not, however, make them better than whites. Quite the opposite, in fact: their physical precocity is proof of their mental retardation! To support this, it is falsely claimed— without references—that infant tests of "intelligence", which measure sensorimotor development (co-ordination between senses and muscles), are *negatively* correlated with IQ. There is, of course, no evidence to show that "baby tests" are negatively correlated with adult IQ. Presumably Eysenck has been influenced by Jensen's reference to the work of Bayley and Schaefer, whose very small-scale study, despite Jensen's misrepresentation, did *not* detect any significant correlation, positive or negative, between "baby tests" and later IQ. A massive study by Broman and co-workers in 1975, with sample sizes in the thousands, did in fact report significant (but small) *positive* correlations between Bayley's baby test scores and subsequent Stanford-Binet IQ. The positive correlations were found in both black and white samples.

IQ and the bones

To a theorist of Eysenck's persuasion, it is important to show that IQ is closely linked to biological and physical measurements. Thus we find Eysenck informing his readers of "the interesting discovery by Sanderson et al (1975) of a marked relationship between intelligence and the shape of the jaw bone, suggesting certain genetic links . . ." (1979). This time we are given a reference, and we are asked to believe that a "marked relationship" has been discovered between intelligence and the shape of the jaw bone, which indicates that the same genes influence both IQ and jaw bone shape. We are not told, however, that the basic finding of Sanderson was that a group of institutionalised persons classed as mentally retarded had peculiar jaw-bone shapes, nor that, as Sanderson pointed out, their jaw shape might be attributable to the periodontal disease that so often occurs in the institutionalised.

Three junior passengers at Johannesburg Airport. Who's got the brains?

Professor Eysenck has also detected a remarkable relationship between IQ and the size of the skull bone. Referring to the work of Van Valen, Eysenck writes: "There is a highly significant correlation between brain size and intelligence, although the absolute value of this correlation is only about 0.3. (Possibly it would be higher if better methods of measuring brain size could be devised.)" (1979).

Ninefold inflation

The Van Valen paper briefly reviews eight studies, most of them done in the 1920s, and most using subjective guesses about intelligence, not actual IQ scores. The estimates of intelligence were correlated with external skull measurements, not with brain size. The "observed correlation" in these studies, according to Van Valen, was a trivial 0.1—not the 0.3 claimed by Eysenck. Van Valen hazarded a guess that, if better measurements were available, the correlation might rise from 0.1 to 0.3. We might hazard the guess that, with better and less subjective measurements, the correlation might sink to 0.0. Either way, Eysenck clearly, and falsely, states that the correlation found was 0.3—accounting for nine times as much of the variance as the actual correlation of 0.1. Eysenck's gloss on this ninefold inflation is to suggest that, with better measurements, the correlation would be still higher.

Civilisation shows signs of surviving

Eysenck's record as a prophet has not been one of unbroken successes. Reviewing the eugenic papers of Burt and Thomson in 1948, Eysenck summarised their argument as follows:

"(1) Intelligence is largely an innate quality. (2) People with lower intelligence tend to have more children than people with higher intelligence. (3) The intellectual level of the population is decreasing in consequence of these facts at a rate of 1.5 to 2 points of IQ per generation. . . . The consequences of such a decline are shattering in their implications. . . . As Burt shows, 'if the rate [of loss] assumed continues, then in little over fifty years the number of pupils of "scholarship" ability would be approximately halved, and the number of feeble-minded almost doubled.' It is doubtful if civilisation as we know it could survive such a catastrophe . . . and there would be very few psychologists nowadays who would not be found in agreement with Burt and Thomson in this matter."

Professor Eysenck joined Burt and his fellow IQ testers in calling for a large-scale study to measure the precise dimensions of the catastrophe which, in the absence of eugenic measures, was overtaking civilisation. A massive study known as the Scottish Survey was then conducted— only to reveal that the national IQ had *increased* by a couple of points over a generation. When engineers build bridges that collapse they are fired, and perhaps tried for negligence. When hereditarians do so, they go on to write new, if repetitive, books.

The matter of social mobility

The persistence of Professor Eysenck, if not his accuracy, is much to be admired. When he discussed IQ and social mobility in 1973, he drew "rather heavily" on the studies of Burt, citing the "outstanding quality of the design and the statistical treatment in his studies". The Burt paper of 1961 was cited to demonstrate that children with high IQs tend to rise in social class, while those with low IQs tend to fall. "The correlation between social mobility and intelligence works out at .38," Eysenck wrote.

When he wrote about social mobility again in 1979, Burt's outstanding designs and statistical treatment lay in ruins. The mathematical implausibility of Burt's too-perfect data on social mobility had been spelled out in devastating detail by Dorfman—sufficiently clearly, in fact, to elicit a concession from Eysenck. However, Eysenck repeated the claim about IQ and social mobility, invoking as evidence this time a small study by Waller:

"The role played by IQ was demonstrated more directly by Waller by correlating father–son differences in IQ with those for occupational status, obtaining . . . a highly significant correlation. . . . Clearly IQ differences play an important part in the process of occupational mobility."

The exact language used in describing this research is of considerable importance. Burt claimed to demonstrate that sons rose above or fell below their parents in occupational level depending on whether they had higher or lower IQs than their fathers. Waller did not present such data, and the claim by Eysenck that Waller had correlated IQ differences with "occupational status" is false. Though Waller had collected data about occupational levels, the correlation was between IQ differences and differences on the Hollingshead Index of Social Position, which combines into a single number the educational level and occupational level.

The fact that a son with an IQ higher than his father's has a higher Hollingshead Index score may mean nothing more than that high-IQ children go to school longer than their fathers did. There is nothing in Waller's paper to show that high IQ sons advanced to higher *occupations* than their fathers, and it contains some evidence to suggest that this might not have been the case. The average educational level achieved by the sons was very significantly higher than that of their fathers; but the average occupational levels of the two generations were exactly the same. The Hollingshead Index, it might be noted, is so constructed that an army sergeant with a university degree has a higher "social position" than a bank president who dropped out of school. The claim about IQ and occupational mobility made by Eysenck has not yet been demonstrated. A reference to false data has simply been replaced by a false reference to real data.

There are many more instances of slipshod and biased references but it seems pointless to continue. The reader should by now understand that not all the confident trumpetings of hereditarians—or of environmentalists—should be taken seriously. The scientist, given half a chance, seems as likely as the used car salesman to try to slip one over. The lay person can only hope that scientists of different persuasions will expose each other's biases and errors; and an attitude of intelligent scepticism on the part of consumers of science is not inappropriate.

20
IN
CONCLUSION

Throughout these pages I have taken a critical—sometimes a sharply critical—stance. The data on heredity and IQ are, at best, ambiguous. Though some are consistent with the notion that IQ is heritable, others are not. The data consistent with a genetic interpretation seem equally consistent with an environmental interpretation. The plausible environmental interpretations have been ignored or soft-pedalled by behaviour geneticists, which might, I have argued, reflect social and political—as well as just plain professional—bias.

Whatever the "experts" may say, there is no compelling evidence that the heritability of IQ is 80 per cent, or 50 per cent, or 20 per cent. There are not even adequate grounds for dismissing the hypothesis that the heritability of IQ is zero. The evidence is clearly inconsistent with a high heritability.

To make such a statement, however, is not to assert that the heritability of "intelligence" is low. We can make no statement at all about how heritable intelligence, or cognitive capacities, or information-processing abilities, might be. We cannot measure such capacities and abilities. We have only IQ tests—limited in their scope, and clearly dependent on past experience. We cannot state that a person with a low IQ score is generally unintelligent and ineffective, any more than we can state the opposite about persons with high IQ scores.

To reject the futile analysis of IQ test scores, however, is in no sense to deny the importance of biological science—or the relevance of genetics to human behaviour and intelligence. This confusion has, I think, been deliberately encouraged. The criticisms of the very real abuses of behaviour genetics have been "answered" by the claim that the critics— unlike behaviour geneticists—are politically motivated. The critics, it is said, wish to deny the biological basis of human inequality. Presumably, the hereditarians' constant harping on the theme of inequality has no political significance.

There are, of course, some theorists who stress man's social nature, while others stress man's biological roots. This division of emphasis is all to the good; in the long run, it can only further our understanding of mankind's complexity. What is intolerable is the self-proclaimed assertion by some hereditarians that they are more "scientific" than environmentalists, more "objective", and innocent of ideological motives. Too often, the appeal by hereditarians to biological science has been nothing more than a clinging to the skirts of a make-believe biology. Too often, that make-believe biology has served to mask honest-to-goodness racism. We conclude with some examples.

Racism, the darker side of make-believe biology

Together with other duties as a school psychologist, Cyril Burt supervised eye examinations of London schoolchildren. He observed that cases of hypermetropia (far-sightedness) seemed rare among Jewish children, and wrote in 1961: "It is tempting to speculate whether the rarity of hypermetropia is itself the effect of a kind of natural selection. Before the invention of spectacles, the Jew whose living depended upon his ability to keep accounts and read them would have been incapacitated by the age of fifty, had he possessed the usual tendency to hypermetropia." This conjuring up of natural selection, of Darwin and of biological science verges on idiocy. The astonishing ignorance—both of European history and of natural selection—revealed in Burt's make-believe biology requires no comment.

The sublime objectivity of biometrical genetics, and its total commitment to the pursuit of pure truth, have never been phrased more movingly than by one of Burt's teachers, the great Karl Pearson. Pearson founded in 1925 a new journal, *Annals of Eugenics,* now known as *Annals of Human Genetics.* Pearson and Elderton wrote the following in 1925 in a foreword to the first issue of the new journal: "We have no axes to grind. . . . We firmly believe that we have no political, no religious and no social prejudices . . . We rejoice in numbers and figures for their own sake and, subject to human fallibility, collect our data—as all scientists must do—to find out the truth that is in them."

The impact of this claim to saintliness of biometrical genetics is diminished when one flips the page to read the very first article published in the new journal. Written by Pearson and Moul, it was entitled "The problem of alien immigration into Great Britain illustrated by an examination of Russian and Polish Jewish children." With 144 tables and 46 figures in 127 pages, immigrant Jewish children in East London were shown to be inferior to the native English as regards teeth, tonsils, adenoids, visual acuity, cleanliness of hair, body and underwear, conscientiousness and intelligence, to have more TB, heart disease, ear disease and eye disease, and to display a tendency to breathe through their mouths. Professor Pearson may have rejoiced in these numbers and figures for their own sake, but he was not oblivious to their implications for immigration policy. For such poor specimens, he wrote, "there should be no place. . . . They will develop into a parasitic race". He admitted, "Some of the children of these alien Jews from the academic standpoint have done brilliantly", but added: "No breeder of cattle, however, would purchase an entire herd because he anticipated finding one or two fine specimens included in it. . . ."

Pearson, it should be noted, was a truly eminent scientist, who doubtless in good conscience declared himself to be an unprejudiced collector of numbers and seeker after truth. His type does not seem to have disappeared from contemporary science. Pearson, as might be guessed, attributed the inferiority of the alien Jews to genetic and biological factors:

> "In the case of the Russian and Polish Jews there has been more or less continuous oppression, nay a veritable selection. . . . Such a treatment does not necessarily leave the best elements of a race surviving. It is likely indeed to weed out the mentally and physically fitter individuals, who alone may have had the courage to resist their oppressors."

Over the years, Pearson suggested, good Jews, brave Jews, clean Jews would have rebelled against the Tsars, and would have been exterminated. The genes for goodness, bravery and cleanliness would thus have died out in the Jewish race; only the dregs would have survived, clamouring for admission to England.

The absurd and bigoted nature of Pearson's make-believe Darwinism is embarrassingly obvious—but not, evidently, to Burt's student, Professor Eysenck. That lover of numbers and admirer of biological science summarised 500 years of American history in the following language in 1971:

> " . . . the more intelligent negroes would have contributed an undue proportion of 'uppity' slaves, as well as being much more likely to try and escape. The terrible fate of slaves falling into either of these categories is only too well known; white slavers wanted dull beasts of burden. . . . Thus there is every reason to expect that the particular sub-sample of the negro race which is constituted of American negroes . . . has been selected throughout history according to criteria which put the highly intelligent at a disadvantage . . . creating a gene pool lacking some of the genes making for higher intelligence."

The identical structure, and the unprovable and unscientific nature, of Pearson's and Eysenck's appeals to "biology" and "genetics" are remarkable. To have substituted black for Jewish scapegoats in a ridiculous paradigm is not, after all, much of a change. We have made depressingly little progress over 50 years—either in science, or in our humanity.

21
REJOINDER
TO
KAMIN

AGREEMENTS AND DISAGREEMENTS

Before dealing in detail with some of the points raised by Professor Kamin, it may be useful to draw attention to the ground rules of scientific debate. These differ in important ways from legal and other types of arguments, which tend to assume the character of *adversary procedures*. In other words, one side (or its legal representatives) puts forward all the arguments which would seem to be favourable to its cause and attempts to dispute all the arguments put forward by the other side, while exactly the same method is used in turn by the opposition. When all else fails, the rule seems to be to abuse the other side's attorney.

In the words of Karl Popper, the philosopher of science, we might say that scientific argument follows the course of *conjectures and refutations*. There are no "sides" in the adversary sense; all those who take part in the argument want to discover "truth" (as far as that is humanly possible) rather than win an argument. Hence a good critic is the best friend of the scientist who puts forward hypotheses or reports experiments. The critic helps him to see whatever weakness there may be in his arguments and demonstrations, and thus enables him to remedy these weaknesses (if they are remediable), improve his theory (unless it has to be jettisoned), and advance somewhat nearer to the truth both he and the critic seek. This picture is of course somewhat idealised, but it does capture, I think, the essential difference between law and science.

The adversary principle

My main objection to Kamin's presentation is that it is based on the adversary principle rather than the truth-finding principle. He attempts to seek out and deploy only those arguments which are in his favour (or can be construed to be so); he disregards those facts and arguments which go counter to his belief; he even descends to the tactic of abusing the opposition's attorney. A whole section (Chapter 19) of his contribution is devoted to a discussion of the alleged vices and follies of HJ Eysenck; this not only does me too much honour but is also clearly irrelevant. Even if I had often been wrong in the past, a fact I would be the last to deny, this would be quite irrelevant to the force of the facts and arguments with which we are dealing. *Argumentum* in a scientific discussion should always be *ad rem*, not *ad hominem*. Although the temptation to answer Kamin's criticisms in detail, pointing out the way in which quotations are wrenched out of context, misinterpreted and generally abused, is almost irresistible, I will not give way to it, but concentrate rather on factual points directly relevant to the great intelligence debate itself.

The critic plays a valuable part

As a beginning, it may be useful to state a number of points on which Kamin and I would seem to agree. Let me start by saying how valuable Kamin's detailed examination of past research reports has been in unearthing errors, bad design, unjustified conclusions and many other faults which had seldom been pointed out so clearly before. This is an important function of the critic in science, and we are all indebted to Kamin for his long-continued labours in this direction.

It is unfortunate that in clearing the way for new and better research, Kamin has fallen into errors at least as grievous as those made by the psychologists he criticises, as David Fulker, in a well-known and lengthy review of Kamin's book published in the *American Journal of Psychology*, has pointed out in considerable detail. The fact is, of course, that secondary sources are always suspect and may be prejudiced, careless or downright wrong; this generalisation applies to all researchers, including Kamin and myself. On any point of factual or theoretical difference, therefore, the reader should ideally go back to the original sources, read them carefully, and then make up his own mind.

Science is always advancing

Readers not widely read in science may be astonished to hear that many of the published accounts contain faults of design, errors of statistical analysis and mistaken interpretations; scientists wise in the practice of their craft will be less likely to be surprised. Methods of experimentation and analysis are constantly improving, and quantitative estimates become more and more accurate. But this does not mean that the earlier work was not "scientific" or important in its own time. For instance, Hubble's constant, the cornerstone of modern cosmology, has changed in size over the years by an order of magnitude. This does not mean the earlier estimates were "unscientific". Science constantly changes and advances; only in the popular imagination does it attain the status of absolute truth.

As social scientists we are subject to restrictions in designing our research. Ideally we would want to know, for instance, the IQs of adopted children, of their adoptive parents, and of their true parents, as well as the principles used by adopting agencies in placing children. But much of this information is not available. So we need to make some assumptions, and these assumptions are of course often subject to debate. Kamin and I would, I presume, agree that such assumptions should, where possible, be tested, and that in any case they should not be derived from *a priori* views about the genetic or environmental origins of IQ.

We would also agree in rejecting early arguments as to the social consequences of IQ testing which were based on scientifically worthless evidence. At around the time of the First World War, a mindless and doctrinaire hereditarianism was common among some psychologists; and an equally mindless and doctrinaire environmentalism later dictated the actions and crimes of such men as Stalin's protégé Trofim Lysenko. Kamin and I are presumably agreed that to make social and political pronouncements based on prejudice, and without solid experimental backing, is fundamentally wrong, and we would both deplore much of what passed for science in those days.

The aim of research in the field of IQ testing should be that of helping the deprived, not enforcing discrimination. Kamin fails to point out in his presentation that the original aim of Godfrey Thomson, Cyril Burt and other British psychologists in introducing IQ tests into the school selection programme was to enable deprived working-class children of high ability to receive a good education. When the Labour government abolished the use of tests, the percentage of working-class youths who went to better-class schools dropped drastically.

Perhaps Kamin and I can agree, too, that IQ testing can have beneficial results as well as harmful ones, and that it is our task as citizens in a democratic society to ensure that the use of such tests (as of all scientific inventions) enhances rather than destroys a society which, however imperfectly, seeks to help the deprived, the poor, and the underprivileged.

ESN classes: a benign policy

It is worth pointing out in this connection that the ESN (educationally sub-normal) classes Kamin is so scathing about are actually an educationally benign way of handling a very difficult problem. Their purpose is to give special educational help to children incapable of benefiting from ordinary school instruction, with a view to integrating them in due course if possible. Experience has shown that much illiteracy is due to the *lack* of ESN classes, and that properly conducted these classes have an extremely important role to fill.

We would also agree in condemning Burt's misdemeanours. There does not now appear to be much doubt that there are many irregularities in Burt's figures which rule them out of court as far as scientific use in the future is concerned. Fakers and fudgers should have no place in science, but in extenuation we may perhaps recall that the divine Newton himself was not above indulging in such practices, as a recent report by RS Westfall indicates.

Kamin and I are in agreement, finally, on many detailed points of behaviour genetics. We are agreed that heritability is a population statistic that may differ from group to group, and from time to time for any particular group. We are agreed that intelligence tests do not measure innate capacity or potential; the most that can be said is that they suggest estimates of this innate capacity which enable us to make reasonably accurate predictions. We are agreed that environment always combines with heredity to produce individual differences, and that the debate is about the respective weight of these two factors in producing such individual differences. We agree that there are very important differences in IQ between different races, and we agree that direct genetic proof that these differences are not environmentally determined has not yet been achieved and may even be impossible. We are agreed that intelligence is not "fixed" in any final, unchangeable form. It would seem to me that this is quite a list of agreements.

Unscientific tactics

To close this section, let me make one further point. Countless critics, in both the scientific and the popular press, have suggested that my views are wrong because there is no direct genetic test of the hypothesis that racial differences are genetically determined. But as I make clear in the body of this book, I have always agreed with this point, and have myself stressed it on many occasions. This tactic can be very confusing to readers, who assume that when a scientist is criticised because a certain assertion is wrong, he must have made that assertion. This adversary sort of argument should have no part in a genuine scientific debate.

Neither should another, similar, tactic which is sometimes labelled "guilt by association". When Kamin associates the early and rather disreputable eugenicists with the quite different later scientific arguments concerning heritability of IQ, the implication is clearly that, since these early advocates of heritability had views which we would now denounce as racist and contrary to reason, more modern advocates of IQ heritability must share similar social views. Such suggestions, being indirect, are all the more difficult to defend against.

ARGUMENTS AND COUNTERARGUMENTS

In this section I propose to discuss certain fairly general disagreements with Kamin. For a start, Kamin tends to talk as if the point of view I represent were genuinely hereditarian. Many of his arguments seem to take the course: X should happen if intelligence were inherited, but what happens is $X - 1$; consequently intelligence is not inherited, or at least it is not possible to assert that it is. But this argument is clearly mistaken. No one in the last 50 years has denied the importance of environment; what is suggested is that heredity and environment contribute to differences in IQ in the proportion, roughly, of 80 per cent and 20 per cent. This statement is often misunderstood to mean that heredity is four times as important as environment. But we are talking about *variances*, and variances are derived from direct measures of variability by squaring the standard deviation. To arrive at a rough and ready estimate of the relative importance of heredity and environment, we must take the root of the proportion given by the variances, namely the square root of $\frac{4}{3}$, which is 2. In other words, *heredity is twice as important as environment*, not four times. This will ring much truer than the incorrect assertion that the ratio is 4 to 1. It leaves a tremendous lot of variability to be manipulated by environmental influences.

Environment—a woolly concept

Unfortunately, the precise nature of the environmental contribution is not specified by Kamin, or indeed by any other subscriber to his way of thinking. Kamin's account mixes up within-family and between-family differences which are two quite different factors. This is a very important point: sometimes he argues as if the contribution were entirely of the one kind, and at other times as if it were entirely of the other kind. Thus in dealing with identical twins brought up in separation he argues that their great similarity is due to the fact that their socio-economic environments (between-family differences) were not so very different, neglecting the fact that their within-family environments were entirely different. When arguing about the differences between MZ and DZ twins, Kamin is entirely concerned with slight discrepancies in upbringing which would fall under the heading of within-family environment and disregards the absence of between-family environmental variance. Nowhere does he make this change of stance clear.

Where is Kamin's theory?

Nor is there any attempt to offer a quantitative estimate of just what contribution each set of factors makes. My own view is that the estimates would be incompatible with each other, but Kamin does not state any alternative theory which could be tested. The theory I present is clear-cut, quantitative and could be disproved directly in various ways. Using Popper's criterion, this makes it a scientific theory. Kamin's failure to formulate a testable, consistent theoretical model of environmentalist influences is perhaps his weakest feature from the scientific point of view. From the point of view of adversary debate, of course, it is a strength. Being absolutely vague, his work escapes test and criticism; if one formulation is found wanting, another can be put in its place; different and incompatible criticisms of the hereditarian stance can be based on different and incompatible models. This makes debate so difficult: one never knows what is being stated positively by environmentalists.

Kamin fails to see the wood for the trees. He always criticises inadequacies, errors, faults and mistakes in individual researchers. This is a necessary task in science. But it is not sufficient to produce general conclusions, which requires looking at all the research literature in the light of the theory under investigation. Nothing less will do.

Important issues are ignored

In failing to take the whole picture into account Kamin sometimes completely passes over relevant evidence contrary to his position. For example, on the subject of regression to the mean, readers will remember that genetic theory predicts, and experiment verifies, that physical and mental traits, including intelligence, show regression to the mean. This fact poses enormous difficulties for any environmentalist explanation of individual differences in IQ, because it is the children born to the most able and successful who regress *downwards* towards the mean, in spite of the environmental advantages their family offer, while it is the children of the dullest and least successful who regress *upwards*, in spite of their deprived upbringing. I know of no attempt to explain this phenomenon in environmental terms. Kamin dismisses the phenomenon as if it had never existed.

Another example is his discussion of the Lawrence orphanage study. I have admitted that too much may have been made of these findings. But the Warsaw study, in which educational and other opportunities were equalised, demonstrates precisely the same effect—namely at best a quite small reduction in the observed differences between children's IQs. Kamin must have been familiar with this study; why did he not mention it?

Yet another example is his treatment of the literature on non-cognitive tests of intelligence, such as Jensen's recent work on reaction times, and the work on EEG evoked potentials. Kamin mentions in passing some early studies of reaction times, but does not discuss the more recent work. Evoked potentials are not mentioned at all. The well-documented correlations between IQ test scores and physiological and behavioural patterns cry out for an environmentalist explanation if Kamin's position is to be at all persuasive, but he passes over the issue without even mentioning the difficulties it raises for his theories.

To return to the subject of regression, it offers in itself impressive proof of the contribution of genetic factors, but its real importance lies in the way that it corroborates the figures arrived at by other means. Estimates of heritability from twin or adoption studies used as a basis for calculating regression effects, predicted very accurately the IQs of the offspring of the very bright children in Terman's large, well-executed study. It is this quantitative agreement between estimates of heritability derived from wholly different approaches which will impress scientists more than any individual approach, with its inevitable assumptions and weaknesses.

Kamin never even looks at the quantitative argument, does not discuss its importance and relevance, and does not warn the reader that here there is a series of facts which any purely environmentalist theory would have the greatest difficulty in dealing with. This is the adversary approach rather than the scientific one.

SOME CRITICISMS ANSWERED

Philosophers of science such as Popper, Kuhn, Lakatos and Suppe are agreed that there are no theories which do not produce anomalies when their assumptions and predictions are tested. The fact that Kamin has discovered a number of anomalies in the literature should be understood for what it is—nothing more than an indication that there are areas where further research which is better designed, better controlled, and on a larger scale is urgently needed. The same is true of the environmentalists' evidence: Kamin himself mentions the contradictory findings of the Texas and Minnesota adoption studies, for example.

Why Kamin's argument won't stand up

Let me now turn to a consideration of some major criticisms made by Kamin of twin studies—studies which bear a considerable burden of the proof adduced by geneticists. Kamin criticises the studies of MZ twins brought up in isolation from each other, largely on the grounds that the socio-economic status of the two families involved was not as dissimilar as it would have been had the twins been assigned to homes on a random basis. This is a simple statement of fact; is it fatal to the argument? Let us look at it in connection with another criticism Kamin makes, this time of the MZ–DZ comparisons which show that MZ twins are much closer to each other on IQ tests than are DZ twins. Here he points out, again quite rightly, that MZ twins are usually treated more alike by their parents than are DZ twins; he argues that this similarity of treatment is responsible for the greater similarity in IQ of the MZ twins.

I have already pointed out that Kamin uses the term "environment" in two distinct senses—to mean between-family environmental differences and within-family environmental differences. But his whole argument is also marred by contradictions. If within-family influences are so strong as to produce the very large differences in IQ between DZ twins, as compared with MZ twins, why are they so impotent in connection with MZ twins reared apart from each other, in different families? If between-family influences are as potent as Kamin suggests in the case of MZ twins brought up in separation, why are DZ twins brought up in the same environment and the same family not more similar? Kamin achieves some plausibility simply by advancing his arguments in different sections of his presentation, and never bringing them together; had he done so the incongruity—the statistical errors and logical confusions—would have been obvious.

Let us look at some figures. Like-sex DZ twins brought up in the same home correlate about .50. MZ twins, brought up in separation, correlate about .75 (ignoring Burt's data, of course). Now for both within-family and between-family environmental factors the DZ twins are clearly more similar; they are brought up in the same home (within-family variance) and as far as between-family variance is concerned the homes of the MZ twins, even if not as dissimilar as chance would decree, are nevertheless dissimilar to some extent, while the DZ twins are of course brought up in one and the same home. Using Kamin's own argument, why is there still this wide discrepancy in favour of the MZ twins when all the environmental factors are more similar for the DZ twins? It is by failing to bring together the relevant figures that Kamin achieves an apparent success with his argument; when we realise the differential meanings of the simple term "environmental" in his two sections, and bring into the argument the appropriate groups, the whole criticism is shown to be based on statistical error and logical confusion.

There are many other reasons for considering Kamin's criticisms invalid. Siblings reared apart only correlate 0.30 at most (disregarding Burt's data). Why are they so much less alike than MZ twins reared apart? As to MZs and DZs, the way they are treated by their parents is much less important than Kamin would have us believe. Sandra Scarr found that whether parents were right or wrong in their classification of twins made little difference to IQ patterns. Loehlin and Nichols looked at the connection between twins' IQ similarity and similarities in the way they were treated; they found no relationship. And does anyone seriously suggest that such factors as dressing twins alike can have any real effect on their intelligence? The whole argument is preposterous, and does not stand up to examination for one minute. It only appears reasonable because Kamin (1) eschews proper statistical evaluation of the data, (2) uses the term "environment" in different and contradictory senses, (3) fails to bring together the arguments used in these two cases, and (4) omits to mention relevant and indeed decisive data. Unfortunately this example of his mode of reasoning is typical of the whole presentation; to criticise every error in it would take several times the length of this book.

A basic false assumption

One important aspect of Kamin's critique is his failure to take a proper quantitative look at the evidence. Fulker, in his review of Kamin's book, lists many almost incredible statistical errors. Here, let us consider as an example the Schiff study which Kamin quotes, in his fourth chapter, as apparently disproving the genetic hypothesis. In brief, the Schiff study found that adopted children had IQs 16 points higher than their stay-at-home siblings; the adoptive homes were all in the upper socio-economic status group. Kamin never asks the only relevant question, namely whether this observed difference is quantitatively compatible with the genetic hypothesis that 80 per cent of the total variance in IQ differences is due to genetic causes. The hypothesis, after all, does admit the importance of environmental factors, indeed that they are half as important as genetic factors. This leaves a wide margin into which the observed differences might fit quite easily.

Let us look at the problem quantitatively. The adoptive homes were about 1.5 standard deviations (SD) above average in socio-economic status, and the native homes about 1 SD below. (In the field of IQ, one SD equals 15 IQ points). The gain in IQ was about .77 SD on the relatively culture-fair ECNI test used. Thus each SD of socio-economic status difference yields a gain of .31 SD for the adopted children (4.6 IQ points). This is well within Jensen's estimate that each SD of environmental variance is worth .43 SD of IQ, or 6.5 IQ points. Thus the figures given by Schiff (which incidentally contain some peculiarities which I have no space to enter into here) do not in any way contradict the hypothesis that IQ differences are produced by genetic factors to the extent of 80 per cent. The same quantitative approach should have been used (but was not) by Kamin in all his comparisons. His argument throughout is based on the erroneous assumption that the genetic hypothesis requires almost complete absence of environmental determination. In that form the genetic hypothesis would clearly be untenable, but of course that is not the form in which it is offered. Kamin is thus guilty not only of *suppressio veri*, but also of the sin of *suggestio falsi*.

As already pointed out, Kamin is very selective in his choice of examples, usually picking those which suit his thesis, and disregarding those that do not. That this is not accidental but intentional becomes clear in his discussion of familial (kinship) relations; he refers to the "artificial median correlations" as if there were some objection to taking many different studies of the same phenomenon and averaging the results to obtain a better estimate of the "true" value to be inserted in the equations. It is difficult to see the basis for such an objection. Different investigators use different samples, different tests, work in different countries, employ different bases for selection. The most representative figure, surely, must be one which takes into account all the studies. It is difficult to think of any other field of science in which it could be seriously argued that there was something wrong and "artificial" in taking into account all the evidence produced by reputable workers in the field.

Other inaccuracies

Kamin cannot be trusted to be factually accurate. To give but one example, he states that the tests used by Shields "were not, unfortunately, well standardised IQ tests". This gives entirely the wrong impression. The Dominoes test was very widely used in the British Army during the war, and was standardised on a larger and more random sample of the population than most other tests in existence. The same applies to Raven's Mill Hill vocabulary test, which Shields also used. Having suggested the use of these tests to Shields as being more suitable for his particular purpose than possible alternatives, I feel that Kamin's derogatory remarks are factually incorrect and motivated perhaps by a desire to impugn, by suggestion, a research project the results of which are difficult to account for in terms of a purely environmentalistic hypothesis. This suggestive approach characterises his whole contribution to this book and it makes it very difficult for the lay reader to disentangle fact from fiction.

Kamin also has a tendency to make statements which contradict the figures he quotes. Thus in making the point that variability in IQ is lower for adoptive parents than in the general population, Kamin talks about expecting "very little variation in IQ among the adoptive parents". Actually the reduction in variation is only from 15 to 11; a standard deviation of 11 is hardly "very little variation". It would be accurate to say that there was a slight reduction in variability amounting to only 4 points of SD. Unwary readers might easily succumb to the implication of Kamin's words and neglect to inspect the figures in order to discover what actually happened.

Another tactic Kamin commonly uses is to talk about things that "probably" happened, or were "likely" to be so. Thus in writing of adoption studies, he states that "agency workers *probably* believe that IQ is largely determined genetically"; that "the agency *may* know the IQ of the unmarried mother"; that the agency will "*perhaps*" know the educational level of the putative father (my italics). In my experience none of these probabilities or possibilities accords with actual agency beliefs and practices, but in any case no argument can be based on a series of surmises without proof of any kind.

It is impossible, for reasons of length, to go through Kamin's whole presentation with a fine-tooth comb, pointing out all the fundamental errors, mistaken assumptions, erroneous statistics, invalid arguments, and downright falsehoods. There is only one bit of advice I can give the reader: Beware! Plausible as Kamin's argument may seem at times, it is built on quicksands, and only examination of the originals and knowledge of the whole literature can save the reader from following the false trails so invitingly laid out for him.

THE ETHICAL PROBLEM

A major difference between environmentalists and so-called hereditarians (who should of course more properly be called interactionists) is their view of the ethical consequences of the empirical findings. The argument is usually advanced that even if what I have said in my presentation is *true*, it is socially *undesirable* that I or anyone should say it, or that further research should be done in this field. It is always unfortunate when one side arrogates to itself an exclusive claim to ethical excellence, and accuses the opposite side of callousness, lack of social sensitivity and immorality. It is particularly when ethical questions get tangled up with political preconceptions that passions are brought into play that should have no place in a scientific controversy.

Social policies should be based on fact

Let me state categorically that the fact that heredity is twice as important as environment in determining differences in intelligence in our type of society cannot be used as an argument against improving social conditions. The evidence is quite clear-cut that such improvement will raise IQ levels considerably, particularly among the deprived, and I can see no rational argument whatsoever to oppose such a course. What

IQ testing does is to pinpoint the groups and the people particularly in need of help, and to monitor whether any of the methods of improvement adopted do in fact have the effects expected of them. This seems to me an entirely benign use of IQ testing, and I can see no ethical objections to it.

When Jensen drew attention to the simple fact that the Headstart movement had largely failed, he was not opposing *realistic* attempts to improve the achievement level and IQ of deprived children; he was concerned to point out that the particular methods adopted by Headstart, and the theories at the basis of the whole operation, were not in line with modern knowledge. Many psychologists, including Jensen and myself, had predicted that Headstart would fail; this does not mean that we would be opposed to more realistic attempts along the same lines, based on proper scientific theories and knowledge. The work of Jensen himself (1972, 1973), Bennett (1976) and Rutter (1979) marks a beginning in the field of rigorous educational research.

The epicentre of the storm over IQ testing has been the racial issue. Are we entitled to label a whole group of people "inferior" on the basis of some form of mental measurement? Can we justify the blow this would be to their self-confidence, their pride and their racial identification? The answer is no, of course, but an additional answer would be that neither Jensen nor I, nor any other responsible psychologist has ever said anything of the kind. What we have pointed out is that group differences, where they exist, hide a good deal of overlap, and that the existence of this overlap makes it absolutely impossible to use race or social class as an index of intelligence, achievement or competence. Each person has to be treated as an individual, and assessed by means of objective criteria.

Our duty to report

It is often argued that while what responsible psychologists say about racial and class differences may be acceptable, the facts and arguments can easily be abused by racists for their own purposes. This is undoubtedly true, but it raises very difficult problems. Should acknowledged facts and correct arguments be kept from the people who pay their salaries by psychologists or other scientists? Should the scientist set up as a censor to keep knowledge from the people? Is there any evidence that racists would be any less extreme in their attitudes if knowledge of such facts as there may be on this issue were to be kept from them? Is their propaganda any more effective because of what science has discovered? There are no certain answers to any of these questions, but it should not be assumed that those who feel that they have a duty to society to make known the results of empirical work are guided by less lofty ethical aspirations than those who hold the opposite view. In my experience, many psychologists interested in the study of genetic influences believe that the obvious social problem produced by the existence of racial and class differences in ability can only be solved, alleviated or attenuated by greater knowledge, and that therefore it is ethically indefensible to refrain from acquiring such knowledge and making it available to society.

To illustrate how the acquisition of genetic knowledge can help in the solution of a problem I will have recourse to an example given by Kamin himself, in connection with quite a different point. Phenylketonuria is a mental disorder produced by a single recessive gene which interferes with the metabolism of phenylanaline, producing toxic substances which lead to a rapid deterioration in mental ability. Recognition of the genetic nature of the defect led to a search for biological causes which was successful, and in turn led to the correct method of treatment, which consists of eliminating analine from the baby's food. The example does not prove, as Kamin suggests, that the environment can alter genetic factors: the individual with this particular hereditary defect possesses a metabolic system which makes it impossible for him to metabolise phenylanaline properly, whatever is done by the environment.

Would it have been ethical to have refrained from investigating the genetic properties of the disorder, and instead proceeded along environmentalist lines—stressing better teaching, more books or better food for the children in question on the assumption that all mental abilities are determined by environmental factors? The recognition of genetic differences leads to an investigation of biological factors, a very necessary step in the logical sequence of investigation which may in the end lead to proper control of intelligence and intellectual differences. The work on evoked potentials already mentioned is a first step in this direction. It would in my view be unethical not to work in this field, because only through such work, I am convinced, will we ever get to grips with the real problems of dullness and low IQ.

The toll of misguided egalitarianism

Beliefs in the determination of intellectual differences by genetic or environmental causes have very important social consequences, and when the beliefs are based on false premises these consequences can be quite serious. One consequence of the widely held belief that environment determines intellectual differences and that all men are equal with respect to intellectual endowment has been the acceptance in many European universities of almost any applicant, regardless of ability or background. Most drastic perhaps has been the effect in Italy, where thousands of ill-prepared and ill-equipped students throng the universities, make normal teaching impossible, and promote a detrimental sub-academic atmosphere and level of instruction. Furthermore, many of the students, unable to achieve examination success at any reasonable level, have produced a situation—by threatening professors, and even taking them prisoner until they agree to the students' proposals—where all are given pass marks (or even awarded first class certificates) for their examinations regardless of the quality of their work. This is now true even in many medical faculties, and the results of this lowering of standards will plague the Italian state for many years to come. Is it not the ethical duty of the scientist to speak out against these false hypotheses in the hope that more appropriate action, based on correct premises, will ensue?

Likewise, the dictates of "affirmative action" have led many American universities and businesses to introduce racist quota systems whereby people are employed or granted studentships on the basis of their race or minority status rather than their ability. This system of "inverse discrimination" has led to problems, including the failure of many blacks admitted under these rules to achieve examination success. It is no kindness to encourage a person to spend years on a university course, only to fail him in the end when that failure was clearly predictable in terms of his IQ scores. Interference in social processes the psychological bases of which are still largely shrouded in mystery is likely to lead to disaster. Our only safeguard is rigorous scientific research, carried out without fear or favour, without prejudice, and without preconceived ideological ideas.

In saying all this I do not wish to give the impression that I am at all certain that the side of the argument I have here presented is right, as far as ethics is concerned, and the other side wrong. I am concerned, rather, to point out that the problem of ethical priorities is a very difficult one indeed, and that any assumption of rectitude on one side or the other should be scrutinised carefully. There are obviously points to support either side, and no one but a fool would assume that one side was wholly right or wholly wrong. My own position was not taken without a great deal of thought and soul-searching, and while I still maintain that in an imperfect world it is probably the most defensible one, and the one most likely to lead to the ultimate advancement of deprived groups, I would not put the point forward with any degree of certainty, nor would I deny my respect to those who disagree with me on conscientious grounds.

ENVOY

So far in this book I have confined myself entirely to scientific arguments, statistical demonstrations and factual material. In this final section I would like to present to the reader some examples, taken from actual life, of the background, social and economic, of a number of geniuses whose intelligence will hardly be in doubt. An environmentalist would have to explain why these people, coming from extremely deprived and often almost incredibly underprivileged backgrounds, succeeded so well in their chosen professions and demonstrated such outstanding intellectual ability.

Geniuses against all odds

One such example is Michael Faraday, arguably the greatest physical scientist of the last century. The modern theory of electricity, with all its practical consequences, is due very largely to his efforts, and his name is universally venerated by scientists. Yet he was the son of an itinerant tinker, had practically no schooling, did not have enough money to own any books, and advanced himself entirely by making good use of such very poor resources as he could gain access to. Readers are invited to consult biographical accounts of his life; they may like to try and see what in his environment could possibly have led him to heights of intellectual achievement which are beyond the reach of the hundreds and thousands of privileged university students studying physics today.

Or consider Isaac Newton, arguably the greatest scientist of all time. He came from a family of small farmers, his father died before he was born, and at birth (Newton was a premature child) he was so frail and puny that two women who went to a neighbour's house to get him a tonic expected to find him dead on their return. Newton was educated in the common village school, which would certainly have been vastly inferior in almost all ways to any modern school. What in his background could have been accountable for his genius? Environmentalists have no answer.

The case of George Washington Carver

My third and last example is perhaps the most convincing of all. He is George Washington Carver, a black born in Missouri during the American Civil War, and probably the greatest American biologist of the last century, despite a background which is a catalogue of appalling misfortunes and deprivations.

His father died before he was born, the ailing son of negro slaves in the deep South. His mother was abducted while he was a baby. He was brought up in a poverty-stricken house by whites who were barely literate. He was denied schooling because of his colour and had to piece together the rudiments of an education while performing the most menial tasks. He was constantly hungry, was dogged by ill health and had a severe stammer thought to have been brought on by childhood traumas.

Yet he succeeded in gaining a formal training—a Bachelor of Science degree in agriculture—and went on to change the agricultural and eating habits of the South, and to carry out original research, working in the field of synthetics (one of the first scientists to do so), creating the science of agricultural chemistry and laying the foundations for the United States peanut industry. His discoveries and inventions are legion. He is also remembered as a talented painter and an indefatigable humanitarian. He devoted his life to the advancement of his race and spurned the honours offered as rewards for his genius. When he died in 1943, he was over 80.

Of the tens of thousands of molly-coddled youngsters receiving higher education in the United States today, with all their advantages, none is likely to achieve a tithe of what the self-taught George Washington Carver achieved. Something, one cannot but feel, has gone seriously wrong. If environment is so all-powerful, then how can the worst imaginable environment produce such a wonderful human being and so outstanding a scientist, and how can the best type of environment that money can buy and the top brains in education conceive produce so vast a number of nonentities, with perhaps a few reasonable scientists sprinkled among them? There is no sign of an answer from environmentalists, though genetics provide the beginnings of an explanation.

In brief

In summary, the genetic hypothesis remains essentially unscathed by Kamin's criticisms, although he must be acknowledged for pointing out the weaknesses of certain studies in precise detail. Kamin is right in emphasising the importance of environmental factors, but wrong in seeming to think that the genetic hypothesis does not allow for these factors in its quantitative formulation. Kamin is right in pointing out the restrictions which must be put on any estimates of heritability, but wrong in thinking that geneticists have ever failed to acknowledge these restrictions, or include them explicitly in their statements. Kamin is right in emphasising the importance of social and ethical considerations in dealing with politically sensitive areas but is wrong in believing that his own side has a monopoly of virtue in this respect.

Last but not least, Kamin is entirely wrong in thinking that there is no evidence to support the view that genetic factors play an important part in producing differences in cognitive ability between people. This notion runs counter to all the available evidence, is contradicted by every expert who has done work in the field, and leaves completely unexplained the quantitative agreement found between many different avenues of approach to the problem of estimating the heritability of intelligence. As Cicero said, 2,000 years ago: "Nihil tam absurde dici potest quod non dicatur ab aliquo philosophorum." Which means: "There is nothing so absurd but some philosopher has said it."

References

It would be inappropriate in a book such as this to document all statements in detail; the list of references would be longer than the book itself. Readers who wish to check up on specific statements, or who wish to read further on any of the topics discussed, are referred to my book *The Structure and Measurement of Intelligence* (New York: Springer, 1980), which contains a full list of references. Alternatively, PE Vernon's book *Intelligence: Heredity and Environment* (San Francisco: Freeman, 1980) may be consulted. A Jensen's *Bias in Mental Testing* (London: Methuen, 1979) is a careful examination of experimental work on the many alleged factors that might bias intelligence measurement and a vital source book on the topic.

Racial differences are discussed in JC Loehlin, G Lindzey and JN Spuhler's *Race Differences in Intelligence* (San Francisco: WH Freeman, 1975) and HJ Eysenck's *Race, Intelligence and Education* (London: M Temple Smith, 1971; entitled *The IQ Argument in the US* and published by Library Press, New York). HJ Jerrison has written the *Evolution of the Brain and Intelligence* (London: Academic Press 1973). Also of interest is a recent publication by RT Osborne, CE Noble, and N Weyl, *Human Variation: The Biopsychology of Age, Race, and Sex*

(London: Academic Press, 1978). Fundamental to any understanding of the genetics of intelligence is K Mather and JL Jinks, *Biometrical Genetics* (London: Chapman & Hall, 1971).

In addition to the material surveyed in these books, and the references they contain, there are a few articles referred to in the text which have appeared too recently to find a place in the review literature; references to these are given below. It cannot be overemphasised, though, that readers wishing to form their own opinion will have to go to the original research literature and work through the publications themselves rather than depend on secondary or tertiary sources; it is not always safe to rely on brief summaries of complex studies, even when there is no intent to mislead.

The work on mental speed and reaction times is summarised by Brand in "General intelligence and mental speed: Their relationship and development" and by Jensen in "Reaction time and intelligence" (both in NATO International Conference on Intelligence and Learning, York University, July 16–20, 1979, to be published). The Polish study of the effects of egalitarian housing and schooling on intellectual differences is discussed by Firkowska et al in "Cognitive development and social policy" (*Science*, 1978, 200, 1357–1362).

The effects of sex-linkage on greater male variability in intelligence are discussed by Lehrke in "Sex linkage: A biological basis for greater male variability in intelligence". The effects of assortative mating on intelligence are discussed by Jensen in "Genetic and behavioral effects of nonrandom mating". Both appear in the book *Human Variation* already mentioned.

Gourlay's "Heredity versus environment: An integrative analysis" (*Psychological Bulletin*, 1979, 86, 596–615) is a recent analysis showing that data used by Jencks et al in *Inequality: A reassessment of the effect of family and schooling in America* (New York: Basic Books, 1972) to show a relatively low heritability for intelligence are not, when properly analysed, contrary to the views expressed here.

Two papers by Richard Lynn—"Selective emigration and the decline of intelligence in Scotland" (*Social Biology*, 1977, 24, 173–182) and "The social ecology of intelligence in the British Isles" (*British Journal of Social and Clinical Psychology*, 1979, 18, 1–12)—demonstrate that it is not very meaningful to talk about large groups, such as "whites" and "blacks". Lynn shows that emigration and other factors produce significant changes in the IQ of given white populations.

In addition to the general treatments of racial differences mentioned above, there are references specific to the chapter on race. There is the classic book by AM Shuey, *The Testing of Negro Intelligence* (New York: Social Science Press, 1966), and McGurk's *Race Differences—20 Years Later* (New York: IAAEE, 1978). The latest survey of ethnic and racial differences by R Lynn, "Ethnic and racial differences in intelligence: International comparisons", appears in the book *Human Variation* already mentioned. Jewish intelligence is surveyed by P Vincent in "The measurement of intelligence of Glasgow Jewish schoolchildren" (*Jewish Journal of Sociology*, 1966, 8, 92–108), and by N Weyl in "Jewish scientists in America" (*Midstream*, 1979, 11–19).

Lastly, I have included a list of references specific to my rejoinder. These are: *Teaching Styles and Pupil Progress* (London: Open Books, 1976) by N Bennett; *Genetics and Education* (London: Methuen, 1972) and *Educability and Group Differences* (London: Methuen, 1973), both by AR Jensen; *The Rise and Fall of TD Lysenko* (New York: Columbia University Press, 1969) by ZA Medveder; *Fifteen Thousand Hours* (London: Open Books, 1979) by M Rutter et al; and "Newton and the fudge factor" (*Science*, 1973, 179, No. 4075, 751–758) by RS Westfall.

22
REJOINDER
TO
EYSENCK

Professor Eysenck has for the most part merely repeated claims in this book that he has made many times before. The reader will note that several of Eysenck's misrepresentations, already discussed in my part of this volume, have been carried forward intact. Thus my rejoinder is in large measure contained in my earlier chapters. There are, however, seven points on which I shall comment briefly. In particular, Eysenck's new effort to put women in their place by the use of IQ data requires some response.

1 The myth of greater male variance
Professor Eysenck suggests that the IQ variance among males is a little larger than that among females. He writes:

"Male–female differences in IQ variability may have a genetic basis in sex-linkage. This hypothesis can be tested directly.... Bayley (1966) has provided relevant data. She found a mother–daughter correlation of 0.68, a father–daughter correlation of 0.66, a mother–son correlation of 0.61, and ... a father–son correlation of 0.44. Brother–sister correlations of 0.55 were found. In other words, the order of size of these correlations is precisely what would be expected on the basis of an X-linked trait."

In Eysenck's version of a sex-linked theory of IQ, the father–son correlation should be lower than any of the other three possible parent–child correlations, which should not differ much from each other. And the brother–sister correlation should be larger than the father–son correlation but smaller than the other parent–child correlations. The Bayley results Eysenck refers to (which were in fact taken by her from a much earlier small-scale study of 51 families reported by Outhit in 1933) do in fact fall into this pattern. But Eysenck neglects to inform his readers that *none* of the correlations in Outhit's study differs significantly from any of the others. In her small study, any fluctuations are attributable, of course, to chance.

Table 9 below summarises 11 different sets of parent–child correlations, broken down by sex, together with brother–sister correlations where they are available. These 11 separate sets of results are all more recent than Outhit's 1933 results, and all the samples are larger—in some cases considerably larger—than Outhit's. They also have another feature in common: not a single one of them displays the pattern of correlations demanded by Eysenck's theory. There are considerable fluctuations in correlations both within and between studies, but they are not systematic. What are we to say of a scholar who presents to his readers the one study out of 12 which—by failing to report that the results are not statistically significant—he can make look consistent with his theory?

Table 9. Family correlations from several studies.

STUDY, TESTS USED	MOTHER-DAUGHTER	MOTHER-SON	FATHER-DAUGHTER	FATHER-SON	BROTHER-SISTER
CONRAD AND JONES, 1940 (ALPHA, STANFORD-BINET)	0.50 (141)	0.48 (128)	0.46 (122)	0.54 (110)	0.54 (374)
CONRAD AND JONES, 1940 (ALPHA, ALPHA)	0.60 (117)	0.39 (128)	0.56 (99)	0.42 (97)	0.55 (144)
GUTTMAN, 1974 (PROGRESSIVE MATRICES)	0.39 (119)	0.24 (89)	0.23 (119)	0.36 (89)	
WILLIAMS, 1975 (WECHSLER)		0.36 (55)		0.43 (55)	
SPUHLER, 1976 (PROGRESSIVE MATRICES)	0.47 (81)	0.14 (81)	0.26 (81)	0.22 (81)	0.08 (58)
KUSE, 1977 (WECHSLER)	0.15 (81)	0.38 (80)	0.08 (81)	0.15 (80)	0.35 (178)
PARK ET AL., 1978 (PROGRESSIVE MATRICES)	0.51 (117)	0.25 (103)	0.39 (112)	0.33 (101)	
SCARR AND WEINBERG, 1978 (WECHSLER)	0.40 (120)*	0.41 (120)*	0.34 (120)*	0.44 (120)*	0.41 (120)*
HORN ET AL., 1979 (WECHSLER)	0.35 (76)	0.10 (86)	0.46 (77)	0.39 (85)	
DEFRIES ET AL., 1979 (PROGRESSIVE MATRICES, "EUROPEANS")	0.25 (692)	0.32 (666)	0.25 (685)	0.23 (672)	0.20 (216)*
DEFRIES ET AL., 1979 (PROGRESSIVE MATRICES, "JAPANESE")	0.25 (248)	0.24 (244)	0.20 (237)	0.09 (241)	0.33 (66)*

(Note: The numbers in parentheses are the number of pairs on which each tabled correlation was based. The asterisks indicate number of families tested, rather than number of pairs.)

Professor Eysenck's dispassionate scholarship has again led him—as it so often has before—to conclusions with great social and political significance. Eysenck's truncated survey of research on parent–child correlations provides a scientific reason—genetic sex-linkage—to expect greater male variability in IQ. This difference in variability is said to be "relatively slight", but "important". The "mathematical properties of the normal curve of distribution", would, we are told, lead us to expect 37 per cent more males than females with IQs above 132. "In the really high-IQ range, the difference would be far greater even than that . . . ."

The point of all this is entirely clear: the science of genetics explains why males furnish "far more geniuses in science, the arts, and other pursuits". Feminists might see discrimination in the fact that leading positions in our society—including professorships at the University of London—are mostly held by males; but such paranoia is put down by the iron laws of genetics. There are very few blacks at the top because the *average* black IQ is low. There are very few women at the top because female IQ *variance* is low. Thus is the world made comfortably safe for white males.

Yet the very words in which Eysenck expresses himself in this volume undercut the genetic argument. Figure 28 shows the extraordinary resemblance between the mental processes—involving both verbal and quantitative intelligence—of Eysenck and Lehrke, an earlier writer on similar topics. There is, so far as I know, no genetic relationship between Eysenck and Lehrke; nevertheless, their intelligences seem as similar as those of identical twins are alleged to be. The Eysenck quotation appears on page 43 of this volume. The Lehrke quotation comes from a book called *Human Variation: The Biopsychology of Age, Race and Sex,* edited by RT Osborne, CE Noble and N Weyl and published in 1978 by Academic Press, New York.

Figure 28. Eysenck and Lehrke: Similarity of mental processes in the absence of genetic relationship

EYSENCK, THIS BOOK (pp 43–4)

"Sexual differentiation in higher animals depends on the sex chromosome complement—two X chromosomes for females, and an X and a Y for males. The X chromosome in man is of medium size, containing about 5 or 6 per cent of the genetic material and carrying about the same proportion of genetic information, including known genes affecting every major body system. The Y chromosome, on the other hand, is one of the smallest, and, as far as is known, carries only the genetic instructions for maleness."

LEHRKE (1978):

"Sexual differentiation in higher animals depends on the sex chromosome complement—two X chromosomes for females, an X and a Y for males. The X-chromosome in man is of medium size, containing about 5 or 6% of the genetic material. . . . It seems to carry about that same proportion of genetic information, including known genes affecting every major body system. The Y-chromosome, on the other hand, is one of the smallest chromosomes and, as far as is known, carries only the genetic instructions for maleness."

Fig. 28 continued

"What one would expect, if there are major genes relating to intelligence on the X chromosome, is that the correlations of test scores for mother-daughter, father-daughter and mother-son would be quite similar, because in each case the parent and child have one X chromosome in common. However, correlations between fathers and sons should be lower since they have no X chromosome in common, and brother-sister correlations should be intermediate since they have an X chromosome in common half the time. Bayley (1966) . . . ."

"What one would expect, if there are major genes relating to intelligence on the X-chromosome, is that the correlations of test scores for mother-daughter, father-daughter, and mother-son would be somewhat similar. In each case, the parent and child have one X-chromosome in common. The correlations between fathers and sons should be lower since they have no X-chromosome in common; and the brother-sister correlation should be intermediate since they have an X-chromosome in common half the time. To quote Bayley (1966) . . . ."

"In other words, the order of size of these correlations is precisely what would be expected on the basis of an X-linked trait. . . . Lehrke discusses both the theory and the evidence at length."

"In other words, the order of size of correlations is exactly what might be expected of an X-linked trait."

"On this basis we would expect 37 per cent more males than females with IQs below 68 or above 132."

"On this basis, there would be expected to be 37% more males than females with IQs below 68, and the same would be true for IQs above 132."

"Such a finding has no bearing on the question of who are the more intelligent, men or women. Dr Samuel Johnson, when asked this question, replied: 'Which man? Which woman?' It is difficult to think of a better conclusion . . . ."

"Dr Samuel Johnson said it most succinctly. When asked which were more intelligent, men or women, he replied, 'Which man? Which woman?'"

To be fair to Professor Eysenck, he does give females credit for some accomplishments:

"They write and spell better, their grammar is better, and they construct sentences better. . . . In other species . . . where emotions are indicated by vocalisations, females also show pronounced superiority. But though females are superior in language usage, or verbal fluency, they are not superior in verbal reasoning. . . . When comprehension and reasoning are taken into account, boys are slightly superior to girls. Females are also better at learning by rote. . . . This ability, too, appears to be genetic."

These musings by Eysenck, in 1980, are reminiscent of the earlier speculations of Eysenck's teacher, Cyril Burt, who wrote in 1911:

"The girls were . . . distinctly better at Erasures, at Speed of Reading and Writing, at Association of Words, and at Completing the Sense of a Story. But . . . in the better tests of reasoning, there is little or no difference; the very slight superiority of the girls may perhaps be due to the slightly superior industry and conscientiousness on their part. . . . Women surpass men especially in those sensations which have a high affective value—smell, colour, tone, touch; men surpass women especially in those sensations which have a high intellectual and practical value—movement, weight, brightness, areas felt, lengths and areas seen. . . . Women excel wherever emotions are seen to interfere with higher mental processes. . . . Wherever there are differences in power of reasoning and of attention, these, when well accredited, seem to be slightly in favour of men. . . . In the adult man, the cortex tends to appear more completely organised; and, in the adult woman, the thalamus tends to appear more completely organised. . . . The mental life of man is predominantly cortical; that of woman predominantly thalamic. . . .

"Mendelian principles in man are those furnished by the temperaments of the North European (or Teutonic) race and South European (or Mediterranean) race. . . . Many of the features in which these two races appear to differ innately from one another resemble those in which the sexes differ. Indeed, a fanciful analogy might easily be drawn both as regards physique and as regards temperament between the typical man and the typical Teuton, and between the typical woman and the typical Mediterranean. . . ."

Eysenck's style of scientific reporting bears many resemblances to Burt's. Though Burt at first grants a slight superiority to girls "in the better tests of reasoning", it is attributed to mere industry and conscientiousness. Within a couple of sentences, in any event, "well accredited" (but un-named) studies of reasoning are said to show that males are superior after all, because of the very structure of the brain and nervous system, of the cortex and the thalamus. The laws of Mendelian inheritance, as reflected in the scientific mirror of psychological tests, determine the mental differences between individuals, between the sexes, and between races. Eysenck repeats Burt's and Lehrke's words in such an astonishingly literal way that we should question his claim that it is females who are "better at learning by rote".

2 The myth of fairness to the sexes

Professor Eysenck defends the practices of IQ test-makers with the following argument:

> "On practically all the IQ tests now in wide use men and women have equal average scores. This is sometimes attributed to some kind of chicanery on the part of psychologists. They are said to have selected items in such a way that equal scores are achieved, regardless of whether there might or might not be genuine differences between the sexes. This accusation is false. Tests such as the Matrices tests, the Dominoes and many others were constructed quite irrespective of sex, and were found to give equal scores to boys and girls, men and women."

What a tangled web Professor Eysenck weaves! The claim that the Matrices Test gives equal scores to men and women is an easily demonstrable falsehood. The manual which accompanies the test in fact cites a study by Heron and Chown which indicates that men receive higher scores on this test than do women, by a very substantial margin. Further, in a massive study conducted by Wilson and others the Matrices Test was given to over 3,000 individuals. There was a hugely significant sex difference: at every age over 18, men were superior to women on this supposedly almost pure test of general intelligence.

The important fact is that we cannot say which sex (or race) might be more intelligent, because we have no way of measuring "intelligence". We have only IQ tests. The makers of most IQ tests—as they themselves freely admit—decided in advance to put together a set of test items which would give men and women equal IQ scores. The equal scores are not a fact of nature but an arbitrary decision of the test-makers, who simply "balance off" items which favour one sex or the other.

Sex differences on individual items are sometimes very large. The famous Wechsler test of adult "intelligence", for example, includes the item "At what temperature does water boil?" Turner and Willerman studied 264 couples and found that 70 per cent of the husbands, and only 30 per cent of the wives, could answer this test item correctly. Does it mean that men are more "intelligent" than women? Or that the husbands can do more creative and useful things with boiling water in the kitchen than their wives? The thought occurs that regarding IQ tests as measures of "intelligence" is nonsensical.

3 The myth of school attainment

Professor Eysenck refers to specific data in support of his claim that IQ tests and tests of scholastic attainment measure two quite different things. We are told that "in studies of school achievement, genetic factors are shown to have far less effect on school achievement than on IQ scores." Reference is made to Husen's 1959 Swedish study, in which twins were tested for achievement in arithmetic, writing and history. "The pattern of variation revealed much lower heritability than for IQ . . . ," Eysenck declares.

Eysenck quoted the Husen study in greater detail in his 1979 book, in which he reported, after using a particular formula for his calculations, that the heritability of school attainment averaged 51 per cent. The same formula showed that another 26 per cent of the total variance in school attainment could be attributed to "common environment". These school attainment figures were summarised as follows: "For IQ, the ratio of genetic variation to common environmental is about $3\frac{1}{2}$ to 1. Here the ratio is on average only 2 to 1."

It is important to note that the $3\frac{1}{2}$ to 1 ratio "for IQ" given by Eysenck is his own estimate, based upon an arbitrary set of IQ studies *not* performed by Husen. What Eysenck neglects to tell his readers—both in 1979 and in the present volume—is that Husen reported the results of IQ tests given to the same large sample of Swedish twins whose school attainments he measured and Eysenck used. The formula favoured by Eysenck, and applied by him to the school attainment data, if applied to these IQ results *for the same twins*, yields a heritability of IQ of 40 per cent, and shows that common environment accounts for another 50 per cent of IQ variance! That, of course, flatly contradicts Eysenck's claim that IQ is more heritable than school attainment.

The point, it should be clear, is not that these various numbers are serious estimates of any facts of nature—but they do bear vivid testimony to Professor Eysenck's methods of scholarship. In this volume I have drawn attention to Eysenck's absurd claim that the heritability of school achievement is much less than half that of IQ, and have said that "few, if any, knowledgeable workers in the area" would accept the claim. I now repeat that statement; and I also repeat Eysenck's candid warning that "the reader will have to rely on the general watchfulness of my colleagues to make sure that I have not tried to slip anything over on him".

4 The myth of equal environments

To make the point that an "egalitarian" social policy cannot eliminate genetically determined social class differences in IQ, Eysenck accepts uncritically the statement made by Firkowska and her colleagues in 1978 that, in Warsaw, "inequalities of habitat among its people" have been eliminated. Eysenck tells us: "People of all levels of education and all types of occupation live in apartments that closely resemble each other, shop in identical stores that contain the same goods, and share similar catering and cultural centres." "School and health facilities" are also said to be the same. Perhaps Eysenck has never visited egalitarian Warsaw; if he has, he has kept his eyes and his mind tightly shut. This same wilful obliteration of the real world makes it possible for him to inform us that white psychologists carrying IQ kits to Africa have been able to measure the inferior intelligence of native blacks in Uganda, Tanzania and South Africa.

5   The myth of regression to the mean

Professor Eysenck carries on at length about "the astonishing fact" of regression to the mean. The fact that the children of high IQ parents do not have IQs as high as those of their parents is taken to support a genetic theory. The genetic principle that parental genes are reshuffled in each generation, causing offspring to regress to the mean, led Eysenck, in dedicating a book to his children, to express the hope "that genetic regression to the mean has not dealt too harshly with them".

Professor Eysenck knows better than this. On many occasions he has been reminded that regression is a necessary *statistical* consequence of the simple fact that the correlation in IQ between parent and child is less than perfect. The regression would occur whether or not genetic factors were involved. The reader should understand that the *parents* of high-IQ *children* also show "regression to the mean". That is, their IQs are not as high as those of their children. Professor Eysenck presumably realises that this regression is not caused by parents inheriting their genes from their children. There seems every likelihood that his own children were bright youngsters; whether regression to the mean has dealt too harshly with Professor Eysenck, I leave for the reader to judge.

6   The myth of evoked potentials

Professor Eysenck's most amazing flights of fancy occur in his section on the "biological measurement of IQ". To show that the "intelligence" measured by IQ tests is real, Eysenck wants to demonstrate that IQ is correlated with underlying "psychophysiological mechanisms". This he does by citing recent work, "some of it not yet published". We are shown evoked potentials (EEG "brainwaves"), taken from a 1969 paper by Ertl and Schafer of 10 high-IQ and 10 low-IQ subjects. We are not told that Ertl himself has not been able to repeat these specimen results; nor have others. We are not told that Ertl cited his "massive research data" in promotional literature for a business firm of which he was president. The firm attempted to sell Ertl's "brain wave analyzer" to school systems as a culture-free intelligence test. The cost of Ertl's brain wave analyzer, in 1976, was $8,500—with a "low-cost service contract available thereafter", and with "per test fees negotiable, based on number of children to be tested".

"In our own laboratory Elaine Hendrickson found evidence to support this finding. . . . Dull subjects produced shallower waves. . . . She found that correlations between evoked potential and IQ now shot up to higher than 0.8—in other words correlations between this psychophysiological measure and IQ were as high as those between one good IQ test and another."

These unpublished results are, to say the least, remarkable. They are even laughable. They are not, however, without precedent. As long ago as 1973 Eysenck quoted Elaine Hendrickson's unpublished research as indicating "the 'true' correlation between evoked potential and verbal intelligence" to be "in excess of 0.6, and possibly 0.7". The same article observed: "It is important to add that in some unpublished research from our laboratory, J Rust found very high heritabilities for amplitude and latency of evoked potentials. . . ." In other words, the "unpublished research" from Eysenck's laboratory was said to demonstrate that (a) evoked potentials were genetically determined, and (b) evoked potentials were highly correlated with IQ. Thus, (c) "biological intelligence" was genetically determined.

There is only one thing wrong with this pretty picture. The "unpublished research" has since been published by Rust, in 1975. Working with sample sizes three times larger than Hendrickson's, Rust failed to find *any correlation at all* between evoked potential and IQ. This damning fact was fully known to Eysenck when he wrote about "unpublished research from our laboratory", but in characteristic fashion he failed to mention it. The convenient finding from Rust's work was put together with the convenient finding from Hendrickson's work. The fact that the two then unpublished studies contradicted each other was conveniently ignored.

The last word on preposterous claims for a massive correlation between IQ and evoked potentials goes to an eminent authority:

". . . a thicket of seemingly inconsistent and confusing findings, confounded variables, methodological differences, statistically questionable conclusions, unbridled theoretical speculation. . . . John Ertl, the field's chief innovator, received the brunt of the most highly publicized criticisms. . . . There have also been a number of failures that seem hard to explain . . . quite different, even contrary, results. . . . The directions of correlations also seem to flip-flop . . . . It appears that measurements of this complex phenomenon have not yet been brought completely under experimental control. . . . The state of the art can hardly be regarded at present as more than exploratory. . . ."

The author of these justly critical remarks is no rabid environmentalist; he is Arthur R Jensen, writing in 1980.

7 The myth of the metals

The most innocent—and perhaps the most revealing—of Eysenck's misrepresentations is his reference to Plato's belief in "genetic causes", as exemplified by the fable of the metals. Men fit to be rulers were said to be made of gold, executives of silver, and farmers and workers of a mixture of iron and brass. In 1979 Eysenck called this "the first clear-cut recognition in print of the importance of individual differences in history". He failed to state that Socrates, who created the fable, described it as a convenient lie, useful to help keep the various social classes in their proper places. When Socrates asked Glaucon, "Do you think there is any way of making them believe it?" the reply was, "Not in the first generation, but you might succeed with the second and later generations."

Professor Eysenck, alas, is not the only psychological "authority" to propagate the myth that science has demonstrated IQ to be highly heritable; nor are his methods of scholarship, as we have seen, unusual in this field of endeavour. The generations of man have continued, and the myth has not yet died. In concluding his contribution to this volume, Eysenck has said that this debate "touches on important social issues" and that the problems "should be discussed calmly and rationally". In his words, "It is to be hoped that the debate carried out within these pages may help to define these issues and enable the reader to form his own conclusions." The reader, I hope, will forgive my occasional sharpness of tone. The social issues *are* important; and there comes a time, I think, to call a myth by its proper name.

References

Adams, G, Ghodsian, M, and Richardson, K. Evidence for a low upper limit of heritability of mental test performance in a national sample of twins. *Nature,* 1976, 263, 314–316.

Bashi, J. Effects of inbreeding on cognitive performance. *Nature,* 1977, 266, 440–442.

Bayley, N, and Schaefer, ES. Correlations of maternal and child behaviors with the development of mental abilities: Data from the Berkeley Growth Study.

Monographs of the Society for Research in Child Development, 1964, 29 (6, Serial No 97).

Bayley, N. Developmental problems of the mentally retarded child. In I Philips (Ed), *Prevention and Treatment of Mental Retardation.* New York: Basic Books, 1966.

Binet, A. *Les idées modernes sur les enfants.* Paris: Flammarion, 1913.

Binyon, M. Scepticism over Burt "fake revelations". *The Times Higher Education Supplement* (London), November 12, 1976, p 14.

Brigham, CC. *A Study of American Intelligence.* Princeton: Princeton University Press, 1923.

Broman, SH, Nichols, PL, and Kennedy, WA. *Preschool IQ: Prenatal and Early Developmental Correlates.* Hillsdale, NJ: Erlbaum, 1975.

Burks, BS. The relative influence of nature and nurture upon mental development: A comparative study of foster parent–foster child resemblance and true parent–true child resemblance. *Yearbook of the National Society for the Study of Education* (Part 1), 1928, 27, 219–316.

Burt, C. Experimental tests of general intelligence. *British Journal of Psychology,* 1909, 3, 94–177.

Burt, C. Tests of higher mental processes and their relation to general intelligence. *Journal of Experimental Pedagogy,* 1911, 1, 93–112.

Burt, C. The inheritance of mental characteristics. *Eugenics Review,* 1912, 4, 168–200.

Burt, C. Ability and income. *British Journal of Educational Psychology,* 1943, 13, 83–98.

Burt, C. *Mental and Scholastic Tests* (2nd Ed) London: Staples, 1947.

Burt, C. The evidence for the concept of intelligence. *British Journal of Educational Psychology,* 1955, 25, 158–177.

Burt, C. *The Backward Child* (4th Ed) London: University of London Press, 1958.

Burt, C. The inheritance of mental ability. *American Psychologist,* 1958, 13, 1–15.

Burt, C. Intelligence and social mobility. *British Journal of Statistical Psychology,* 1961, 14, 3–24.

Burt, C. *The Backward Child* (5th Ed) London: University of London Press, 1961.

Burt, C. The genetic determination of differences in intelligence: A study of monozygotic twins reared together and apart. *British Journal of Psychology,* 1966, 57, 137–153.

Burt, C, and Howard, M. The multifactorial theory of inheritance and its application to intelligence. *British Journal of Statistical Psychology,* 1956, 9, 95–131.

Cleary, TA, Humphreys, LG, Kendrick, SA, and Wesman, A. Educational uses of tests with disadvantaged students. *American Psychologist,* 1975, 30, 15–41.

Conrad, HS, and Jones, HE. A second study of familial resemblance in intelligence: Environmental and genetic implications of parent–child and sibling correlations in the total sample. *Yearbook of the National Society for the Study of Education,* (Part 2) pp 97–141. Bloomington: Public School Publishing Co, 1940.

Conway, J. The inheritance of intelligence and its social implications. *British Journal of Statistical Psychology,* 1958, 11, 171–190.

DeFries, JC, Johnson, RC, Kuse, AR, McClearn, GE, Polovina, J, Vandenberg, SG, and Wilson, JR. Familial resemblance for specific cognitive abilities. *Behavior Genetics,* 1979, 9, 23–43.

Dorfman, DD. The Cyril Burt question: New findings. *Science,* 1978, 201, 1177–1186.

Eaves, LJ. Testing models for variation in intelligence. *Heredity,* 1975, 34, 132–136.

Eaves, LJ. Inferring the causes of human variation. *Journal of the Royal Statistical Society, Series A,* 1977, 140, 324–348.

Erlenmeyer-Kimling, L, and Jarvik, LF. Genetics and intelligence: a review. *Science,* 1963, 142, 1477–1479.

Ertl, J, and Schafer, EWP. Brain response correlates of psychometric intelligence, *Nature,* 1969, 223, 421–422.

Eysenck, HJ. Some recent studies of intelligence. *Eugenics Review,* 1948, 40, 21–22.

Eysenck, HJ. *Race, intelligence and education.* London: Temple Smith, 1971.

Eysenck, HJ. *The Measurement of Intelligence.* Lancaster: Medical and Technical Publishing Co, 1973.

Eysenck, HJ. *The Inequality of Man.* London: Temple Smith, 1973.

Eysenck, HJ. HJ Eysenck in rebuttal. *Change,* 1974, 6, 2.

Eysenck, HJ. After Burt. *New Scientist,* 1976, 72, 488.

Eysenck, HJ. The case of Sir Cyril Burt. *Encounter,* 1977, 48, 19–24.

Eysenck, HJ. Sir Cyril Burt and the inheritance of the IQ. *New Zealand Psychologist,* 1978.

Eysenck, HJ. *The Structure and Measurement of Intelligence.* New York: Springer-Verlag, 1979.

Fabsitz, RR, Garrison, RJ, Feinleib, M, and Hjortland, M. A twin analysis of dietary intake: Evidence for a need to control for possible environmental differences in MZ and DZ twins. *Behavior Genetics,* 1978, 8, 15–25.

Feldman, MW and Lewontin, RC. The heritability hang-up. *Science,* 1975, 190, 1163–1168.

Firkowska, A, Ostrowska, A, Sokolowska, M, Stein, Z, Susser, M and Wald, I. Cognitive development and social policy. *Science,* 1978, 200, 1357–1362.

Fisher, RA. The correlation between relatives on the supposition of Mendelian inheritance. *Transactions of the Royal Society of Edinburgh,* 1918, 52, 399–433.

Freeman, FN, Holzinger, KJ and Mitchell, BC. The influence of environment on the intelligence, school achievement and conduct of foster children. *Yearbook of the National Society for the Study of Education* (Part 1), 1928, 27, 103–205.

Fulker, DW. Review of *The Science and Politics of IQ,* by LJ Kamin. *American Journal of Psychology,* 1975, 88, 505–519.

Fuller, JL and Thompson, WR. *Foundations of behavior genetics.* St Louis, Missouri: Mosby, 1978.

Galton, F. *Hereditary Genius: An Inquiry into Its Laws and Consequences.* London: Macmillan, 1869.

Goddard, HH. The Binet tests in relation to immigration. *Journal of Psycho-Asthenics,* 1913, 18, 105–107.

Goddard, HH. *Human Efficiency and Levels of Intelligence.* Princeton: Princeton University Press, 1920.

Goldberger, AS. Heritability. *Economica,* 1979, 46, 327–347.

Guttman, R. Genetic analyses of analytical spatial ability: Raven's Progressive Matrices. *Behavior Genetics,* 1974, 4, 273–283.

Hearnshaw, LS. *Cyril Burt: Psychologist.* Ithaca: Cornell University Press, 1979.

Herrman, L and Hogben, L. The intellectual resemblance of twins. *Proceedings of the Royal Society of Edinburgh,* 1932, 53, 105–129.

Herrnstein, RJ. Kamin errs . . . Herrnstein. *Contemporary Psychology*, 1975, 20, 758.

Herrnstein, RJ. *IQ in the meritocracy*. Boston: Atlantic Monthly Press, 1973.

Horn, JM, Loehlin, JC and Willerman, L. Intellectual resemblance among adoptive and biological relatives: The Texas Adoption Project. *Behavior Genetics*, 1979, 9, 177–207.

Huntley, RMC. Heritability of intelligence. In JE Meade and AS Parkes (Eds), *Genetic and Environmental Factors in Human Ability*. New York: Plenum, 1966.

Husen, T. *Psychological Twin Research: A Methodological Study*. Stockholm: Almquist and Wiksell, 1959.

Jencks, C. *Inequality*. New York: Basic Books, 1972.

Jensen, AR. How much can we boost IQ and scholastic achievement? *Harvard Educational Review*, 1969, 39, 1–123.

Jensen, AR. *Bias in Mental Testing*. New York: Free Press, 1980.

Jensen, AR. Sir Cyril Burt (obituary). *Psychometrika*, 1972, 37, 115–117.

Jensen, AR. IQs of identical twins reared apart. *Behavior Genetics*, 1970, 1, 133–148.

Jensen, AR. Kinship correlations reported by Sir Cyril Burt. *Behavior Genetics*, 1974, 4, 1–28.

Jensen, AR. Race and mental ability. In FJ Ebling (Ed): *Racial Variation in Man*. New York: Wiley, 1975.

Jensen, AR. Heredity and intelligence: Sir Cyril Burt's findings. Letter to the London *Times*, December 9, 1976, p 11.

Jinks, JL and Fulker, DW. A comparison of the biometrical genetical, MAVA, and classical approaches to the analysis of human behavior. *Psychological Bulletin*, 1970, 73, 311–349.

Juel-Nielsen, N. Individual and environment: a psychiatric-psychological investigation of monozygous twins reared apart. *Acta Psychiatrica et Neurologica Scandinavica*, (Monograph Supplement 183), 1965.

Kamin, LJ. Heredity, intelligence, politics, and psychology. Invited address, Eastern Psychological Association, 1973.

Kamin, LJ. *The Science and Politics of IQ*. Potomac, Maryland: Erlbaum, 1974.

Kamin, LJ. Comment on Munsinger's adoption study. *Behavior Genetics*, 1977, 7, 403–406.

Kamin, LJ. Transfusion syndrome and the heritability of IQ. *Annals of Human Genetics*, 1978, 42, 161–171.

Kamin, LJ. Psychology as social science: The Jensen affair, ten years after. Presidential address, Eastern Psychological Association, 1979.

Kamin, LJ. Inbreeding depression and IQ. *Psychological bulletin*, 1980, 87, 434–443.

Kuse, AR. *Familial resemblances for cognitive abilities estimated from two test batteries in Hawaii*. Unpublished doctoral dissertation, University of Colorado, 1977.

Lawrence, EM. An investigation into the relation between intelligence and inheritance. *British Journal of Psychology*, (Monograph Supplement 5), 1931.

Layzer, D. Heritability analyses of IQ scores: Science or numerology? *Science*, 1974, 183, 1259–1266.

Leahy, A. Nature–nurture and intelligence. *Genetic Psychology Monographs*, 1935, 17, 241–306.

Lehrke, RG. Sex linkage: a biological basis for greater male variability in intelligence. In RT Osborne, CE Noble and N Weyl (Eds), *Human Variation:*

The Biopsychology of Age, Race, and Sex. New York: Academic Press, 1978.

Lippmann, W. The abuse of the tests. In NJ Block and G Dworkin (Eds), *The IQ controversy.* New York: Pantheon, 1976.

Loehlin, JC. Combining data from different groups in human behavior genetics. In JR Royce (Ed): *Theoretical Advances in Behavior Genetics.* Leiden: Sijthoff and Hoordhoff, 1979.

Loehlin, JC., Liudzey, G and Spuhler, JC *Race Differences in Intelligence.* San Francisco: Freeman, 1975.

Loehlin, JC and Nichols, RC. *Heredity, Environment and Personality.* Austin: University of Texas Press, 1976.

Mackintosh, NJ. Review of *Cyril Burt: Psychologist,* by LS Hearnshaw. *British Journal of Psychology,* 1980, 71, 174–175.

Munsinger, H. Children's resemblance to their biological and adopting parents in two ethnic groups. *Behavior Genetics,* 1975, 5, 239–254.

Munsinger, H. A reply to Kamin. *Behavior Genetics,* 1977, 7, 407–409.

Munsinger, H. The identical-twin transfusion syndrome: A source of error in estimating IQ resemblance and heritability. *Annals of Human Genetics,* 1977, 40, 307–321.

Newman, HH, Freeman, FN and Holzinger, KJ. *Twins: A Study of Heredity and Environment.* Chicago: University of Chicago Press, 1937.

Nichols, RC. The inheritance of general and specific ability. *National Merit Scholarship Research Reports,* 1965, 1, 1–9.

Outhit, MC. A study of the resemblance of parents and children in general intelligence. *Archives of Psychology,* 1933, No 149.

Park, J, Johnson, RC, DeFries, JC, McClearn, GE, Mi, MP, Rashad, MN, Vandenberg, SG and Wilson, JR. Parent–offspring resemblance for specific cognitive abilities in Korea. *Behavior Genetics,* 1978, 8, 43–52.

Pastore, N. *The Nature–Nurture Controversy.* New York: Columbia University Press, 1949.

Pearson, K and Elderton, EM. Foreword. *Annals of Eugenics,* 1925, 1, 1–4.

Pearson, K and Moul, M. The problem of alien immigration into Great Britain illustrated by an examination of Russian and Polish Jewish children. *Annals of Eugenics,* 1925, 1, 5–127.

Rao, DC and Morton, NE. IQ as a paradigm in genetic epidemiology. In NE Morton and CS Chung (Eds), *Genetic epidemiology.* New York: Academic Press, 1978.

Rao, DC, Morton, NE and Yee, S. Resolution of cultural and biological inheritance by path analysis. *American Journal of Human Genetics,* 1976, 28, 228–242.

Record, RG, McKeown, T and Edwards, JH. The relation of measured intelligence to birth weight and duration of gestation. *Annals of Human Genetics,* 1969, 33, 71–79.

Record, RG, McKeown, T and Edwards, JH. An investigation of the difference in measured intelligence between twins and single births. *Annals of Human Genetics,* 1970, 34, 11–20.

Rust, J. Cortical evoked potential, personality and intelligence. *Journal of Comparative and Physiological Psychology,* 1975, 89, 1220–1226.

Sanderson, A, Laycock, PJ, MacCulloch, H and Girling, A. Morphological jaw differences in mentally subnormal and normal adult males. *Journal of Biosocial Science,*1975, 7, 393–410.

Scarr, S. *IQ: Race, Social Class and Individual Differences.* Hillsdale, NJ: Erlbaum, 1980.

Scarr, S and Weinberg, RA. Intellectual similarities within families of both adopted and biological children. *Intelligence,* 1977, 1, 170–191.

Scarr, S and Weinberg, RA. The influence of "family background" on intellectual attainment: The unique contribution of adoption studies. *American Sociological Review,* 1978, 43, 674–692.

Schiff, M, Duyme, M, Dumaret, A, Stewart, J, Tomkiewicz, S and Feingold, J. Intellectual status of working-class children adopted early into upper middle-class families. *Science,* 1978, 200, 1503–1504.

Schull, WJ and Neel, JV. *The Effects of Inbreeding on Japanese Children.* New York: Harper and Row, 1965.

Shields, J. *Monozygotic Twins Brought up Apart and Brought up Together.* London: Oxford University Press, 1962.

Shields, J. MZ twins: Their use and abuse. In W Nance (Ed), *Twin Research: Psychology and Methodology.* New York: Liss, 1978.

Skodak, M and Skeels, H. A final follow-up study of one hundred adopted children. *Journal of Genetic Psychology,* 1949, 75, 85–125.

Smith, RT. A comparison of socio-environmental factors in monozygotic and dizygotic twins, testing an assumption. In SG Vandenberg (Ed), *Methods and Goals in Human Behavior Genetics.* New York: Academic Press, 1965.

Snygg, D. The relation between the intelligence of mothers and of their children living in foster homes. *Journal of Genetic Psychology,* 1938, 52, 401–406.

Spuhler, KP. Family resemblance for cognitive performance: An assessment of genetic and environmental contributions to variation. Unpublished doctoral dissertation, University of Colorado, 1976.

Stocks, P and Karn, MN. A biometric investigation of twins and their brothers and sisters. *Annals of Eugenics,* 1933, 5, 1–55.

Tabah, L and Sutter, J. Le niveau intellectuel des enfants d'une même famille. *Annals of Human Genetics,* 1954, 19, 120–150.

Taubman, P. The determinants of earnings: genetics, family and other environments. A study of white male twins. *American Economic Review,* 1976, 66, 858–870.

Terman, LM. *The Measurement of Intelligence.* Boston: Houghton-Mifflin, 1916.

Turner, RG and Willerman, L. Sex differences in WAIS item performance. *Journal of Clinical Psychology,* 1977, 33, 795–797.

Vandenberg, SG. What do we know today about the inheritance of intelligence and how do we know it? In R Cancro (Ed), *Intelligence, Genetic and Environmental Influences.* New York: Grune and Stratton, 1971.

Van Valen, L. Brain size and intelligence in man. *American Journal of Physical Anthropology,* 1976, 40, 417–424.

Waller, JH. Achievement and social mobility: relationships among IQ score, education and occupation in two generations. *Social Biology,* 1971, 18, 252–259.

Williams, T. Family resemblance in abilities: The Wechsler scales. *Behavior Genetics,* 1975, 5, 405–409.

Wilson, JR, DeFries, JC, McClearn, GE, Vandenberg, SG, Johnson, RC and Rashad, MN. Cognitive abilities: Use of family data as a control to assess sex and age differences in two ethnic groups. *International Journal of Aging and Human Development,* 1975, 6, 261–276.

amplitude (of brain wave) depth

analysis of variance statistical procedure for analysing the interaction of two or more factors

artifact effect resulting from human, rather than natural, processes

assortative mating tendency for spouses to be genetically similar, for instance in intelligence

behavioural genetics the study of the influence of heredity on behaviour

biometric applying statistics to biological data

bit unit of information

blind study study in which the researcher, to avoid bias, does not know which subjects are receiving which treatment

chromosome large molecules which contain the genes responsible for hereditary traits

cognition mental processes (eg thinking and perception) whereby things are known

control group group of subjects as similar as possible to the experimental group and submitted to all the same conditions except the one being studied

controlled study research study in which important characteristics (eg age, sex and social status) of subjects are taken into account

consanguinity blood relatedness

convergent test item question with a single correct answer

correlation measure of degree of relationship between two factors, expressed as a *correlation coefficient*

covariance tendency for two factors to vary together

culture-fair test test of natural ability in which a person's background is of little importance

crystallised intelligence ability dependent on acquired knowledge

divergent test item question with no single correct answer, designed to test originality

dizygotic (DZ) twins fraternal twins, developed from two ova fertilised by two sperms. They may be of the same or opposite sex

dominance in Mendelian inheritance, the power of one member of a pair of factors to suppress the appearance of the other (recessive) member. High intelligence supposedly has dominance over low intelligence

dominant trait trait which will be expressed in any individual who has its gene (see *recessive trait*)

electroencephalograph (EEG) machine which records brainwaves

environmentalist stressing the importance of the environment as against heredity

eugenics the study of inherited human characteristics, particularly with a view to their improvement

evoked potential brain wave produced by a sudden stimulus.

factor analysis statistical technique to identify the relative importance of factors contributing to a complex ability or trait

fluid intelligence natural ability which is not dependent on acquired knowledge

g term used by Spearman to denote general intelligence

gene carrier of a hereditary factor. Contained in a chromosome

hereditarian stressing the importance of heredity as against environment

heritable which can be inherited

heritability extent to which a trait can be inherited

hypothesis tentative explanation or theory

inbreeding depression lowering of the value of a trait, for instance IQ, as a result of marriage between blood relatives

information processing theory the use of computer programs as a model for the way the mind processes information

IQ intelligence quotient. The ratio of mental age to chronological age (average = 100)

latency (of response) length of time elapsing between stimulus and response

matched groups groups matched for characteristics not under investigation (eg age, ability, education), so that differences in the factor being studied will not be affected by incidental differences (see *control group*)

Matrices type of IQ test, eg Raven's Progressive Matrices

mean arithmetical average

median statistical term for the middle number in a series of numbers arranged in order of magnitude

Mendelian inheritance laws governing the inheritance of characteristics as determined by Gregor Mendel

model theoretical framework developed in one field and applied to another for clarity

monozygotic (MZ) twins identical twins, developed from one ovum and one sperm

neuron nerve cell

normal distribution distribution of a given trait in a large population, represented as a bell-shaped curve. If a trait is normally distributed, most people will cluster around the average

operational definition definition of a concept (eg intelligence, heat) in terms of the methods used to measure it

paradigm model or pattern

parameter agreed limits or scope

primary abilities term used by Thurstone to denote the different abilities (eg verbal, numerical) which make up intelligence

raw data numerical data in their original form, before conversion, analysis or interpretation

reaction time time taken to react to a stimulus in a test

recessive trait trait which will not be expressed if its gene is paired with a corresponding dominant gene

regression to the mean tendency for the offspring of parents who are extreme in a given trait to be closer to the average

sample group selected for study

SES socio-economic status

sex-linkage the association of certain traits with a person's sex

sibling brother or sister

standard deviation measure of variability computed by squaring the root of the mean deviation. With IQ, one standard deviation is 15 points

standardised test a test which has been administered to large samples of people and for which the performance norms of different groups have been established

Stanford–Binet test most widely used children's intelligence test

validity degree to which a test measures what it claims to measure. Internal validity is agreement with other tests which measure the same factor. External validity is agreement with indices other than tests

variability in statistics, the dispersion of values from the average. Standard deviation and variance are measures of variability

variance measure of variability equivalent to the average of the squares of the individual deviations from the mean

WAIS Wechsler Adult Intelligence Scale, an intelligence test

WISC children's version of the Wechsler test for adults